The Pursuit of Happiness

T0225386

The Pursuit of Happiness: Between Prosperity and Adversity looks at activities, practices, and experiences that are instrumental in changing one's level of well-being.

This book focuses on the situations in which well-being is challenged, or even decreased, and explores, guided by *Dialogical Self Theory*, pathways that lead to its elevation. Research has suggested that there are three main determinants of well-being: genetic factors, one's individual's history, and happiness-relevant activities. The third and most promising means of altering one's happiness level are activities and practices that require some degree of effort. A surprising finding is that these personal efforts may have a happiness-boosting potential that is almost as large as the probable role of genetics, and apparently larger than the influence of one's individual history. Efforts are invested in fields of tension between prosperity and adversity.

The Pursuit of Happiness covers a variety of topics, such as finding happiness and well-being in the face of extreme adversity, the role of honesty in genuine happiness, the promise of minimalistic life orientations, the value of inner silence, evaluating our lives from a future perspective, and the relationship between happiness, career development, counselling, and psychotherapy.

This book was originally published as a special issue of the *British Journal of Guidance & Counselling*.

Hubert J.M. Hermans is an Emeritus Professor of Psychology at Radboud University, Nijmegen, the Netherlands. He is a Creator of Dialogical Self Theory and honorary president of the International Society for Dialogical Science.

Frans Meijers was an Associate Professor at The Hague University, the Netherlands, and a Co-editor of the *Journal of Guidance and Counselling*, International Symposium Series.

The Pursuit of Happiness

Between Prosperity and Adversity

Edited by
Hubert J.M. Hermans and Frans Meijers

Routledge
Taylor & Francis Group

LONDON AND NEW YORK

British Library Cataloguing-in-Publication Data
A catalogue record for this book is available from the British Library

ISBN: 978-0-367-43712-1 (hbk)
ISBN: 978-1-03-283899-1 (pbk)
ISBN: 978-1-00-300883-5 (ebk)

DOI: 10.4324/9781003008835

Typeset in Myriad Pro
by codeMantra

Dedication to Frans Meijers

Frans was a man who knew happiness. Through hard work, dedication, service to others, and a strong streak of rebelliousness, he influenced and inspired many people, especially educators in vocational and higher education. In addition to intellectual and theoretical pursuits, he was also fascinated by dialogue and the role of emotions in learning processes. He knew that in addition to good theory and research, the affective dimension would further foster a meaningful conversation and he began exploring this with colleagues Wijers and Wardekker in the 1990s.

Frans often told me that to his detriment, he had ignored his own feelings throughout life. Though they bubbled from him in quite obvious ways, he didn't open easily. He grew up in a loving working-class family where "emotions were there, but not spoken about", and he felt this had limited his growth and happiness. He spoke about how his own struggle had become an impetus to create integrative learning opportunities for youth and adults: he believed what career researcher, Mark Savickas, says about our life and vocational choices, "we are always actively mastering in life what we have passively suffered."

As Frans began to work on research and methods that inspired him and required that he explores his own emotions, his work blossomed. One colleague said, "his theories truly became infused with life." He told me the last ten years of his life were the happiest and vital to his learning. His natural exuberance combined with a new awareness of his feelings allowed him meaningful dialogues and this left him fulfilled. While he was ill, he would even say, "except for the fact that I am dying, I have nothing to complain about." He considered dying a highly social activity that involved expression of feelings and true conversations. Even in this phase of his life, he knew happiness and he brought people joy and inspired them by his example.

Frans worked on many books and journals throughout his life, and he was a pioneer in career guidance education in the Netherlands, later breaking through internationally. Ultimately, he believed dialogues about actual lived experiences were the key to meaning-oriented learning and that if you could create places where this was possible, you would be able to offer safe (enough), challenging, and ultimately powerful learning environments.

This collection of writings is in the spirit of what Frans wanted to share with the world: ways of learning about ourselves and transforming our experiences and challenges through dialogue in a way that will support our happiness and well-being. He would have been proud and delighted to see this book.

Reinekke Lengelle
Edmonton, 24 September 2019

Reinekke Lengelle, PhD, was Frans's life and work partner. She is an Assistant Professor of Interdisciplinary Studies at Athabasca University, Canada, and a Researcher with The Hague University, the Netherlands. She is also a symposium Co-editor with the *British Journal of Guidance & Counselling*.

Contents

Citation Information

The chapters in this book were originally published in the *British Journal of Guidance & Counselling*, volume 47, issue 2 (April 2019). When citing this material, please use the original page numbering for each article, as follows:

Chapter 1
The pursuit of happiness
Hubert Hermans and Frans Meijers
British Journal of Guidance & Counselling, volume 47, issue 2 (April 2019) pp. 139–142

Chapter 2
Finding happiness and wellbeing in the face of extreme adversity
Katrin Den Elzen
British Journal of Guidance & Counselling, volume 47, issue 2 (April 2019) pp. 143–156

Chapter 3
Renewal and multi-voicedness: in search of narrative conscience
Michiel de Ronde
British Journal of Guidance & Counselling, volume 47, issue 2 (April 2019) pp. 157–167

Chapter 4
Minimalist life orientations as a dialogical tool for happiness
Jennifer E. Hausen
British Journal of Guidance & Counselling, volume 47, issue 2 (April 2019) pp. 168–179

Chapter 5
The search for inner silence as a source for Eudemonia
Olga V. Lehmann, Goran Kardum and Sven Hroar Klempe
British Journal of Guidance & Counselling, volume 47, issue 2 (April 2019) pp. 180–189

Chapter 6
Creating space for happiness to emerge: the processes of emotional change in the dialogical stage model
Georgia Gkantona
British Journal of Guidance & Counselling, volume 47, issue 2 (April 2019) pp. 190–199

For any permission-related enquiries please visit:
http://www.tandfonline.com/page/help/permissions

Contributors

Anuradha J. Bakshi is an Associate Professor and Head of the Human Development Specialization at the Nirmala Niketan College of Home Science, University of Mumbai, India.

Margaret P. Boyer is an advanced doctoral student in the Department of Counselling, Clinical, and School Psychology at the University of California, Santa Barbara, USA.

Collie W. Conoley is the Director of the Carol Ackerman Positive Psychology Centre at the University of California, Santa Barbara, USA and a Professor in the Department of Counselling, Clinical, and School Psychology at the University of California, Santa Barbara, USA.

Katrin Den Elzen is a doctoral candidate in creative writing at Curtin University, Perth, Australia. Her research focuses on identity construction in the memoir, particularly the grief memoir.

Graham A du Plessis is a registered Clinical Psychologist (South Africa and New Zealand) and a Psychologist (Australia). He is a Senior Research Fellow in Psychology at the University of Johannesburg, South Africa, and a Lecturer in the Faculty of Business, Economics and Law at the University of Queensland, Brisbane, Australia.

Georgia Gkantona studied Psychology (BSc) at the University of Athens, Greece, and got Master's and Doctoral Degree in Counselling Psychology from the same university.

Tharina Guse is a Counselling Psychologist, Professor, and Head of the Department of Psychology at the University of Pretoria, South Africa.

Jennifer E. Hausen is a Psychologist and Research Associate at the University of Luxembourg. Her primary research interests include subjective well-being and values.

Kathryn J. Hawley is a doctoral candidate in the Department of Counselling, Clinical, and School Psychology at the University of California, Santa Barbara, USA.

Hubert J.M. Hermans is an Emeritus Professor of Psychology at Radboud University, Nijmegen, the Netherlands. He is a Creator of Dialogical Self Theory and the President of the International Society for Dialogical Science.

Michael Huston is a Counsellor and Career Practitioner and has worked in post-secondary counselling and counsellor education for over 25 years. Currently, he is an Associate Professor and Counsellor at Mount Royal University in Calgary, Canada.

Goran Kardum is an Associate Professor in the Department of Psychology at the Faculty of Humanities and Social Sciences at the University of Split, Croatia.

Sven Hroar Klempe is an Associate Professor in Psychology in the Department of Psychology at the NTNU Norwegian University of Science and Technology, Trondheim, Norway.

Olga V. Lehmann is a Psychologist, Poet, Teacher, and Researcher. She is a Postdoctoral Fellow in the Department of Mental Health at the NTNU Norwegian University of Science and Technology, Trondheim, Norway.

Małgorzata Łysiak is an Assistant Professor in the Institute of Psychology, Department of Clinical Psychology at the John Paul II Catholic University of Lublin, Poland.

Frans Meijers was an Associate Professor at The Hague University, the Netherlands, and a Co-editor of the *Journal of Guidance and Counselling*, International Symposium Series.

Evelyn I. Winter Plumb is a doctoral candidate in the Department of Counselling, Clinical, and School Psychology at the University of California, Santa Barbara, USA.

Małgorzata Puchalska-Wasyl is an Associate Professor of Psychology at the John Paul II Catholic University of Lublin, Poland.

Dave E. Redekopp has worked on the development of better career development and workplace concepts and practices for over 30 years. He holds a PhD in Educational Psychology (Theory & Research) and a Master's degree in Educational Psychology (Counselling), both from the University of Alberta, Edmonton, Canada.

Dr. Michiel de Ronde is an Applied Research Professor of Guidance Studies at the Research Centre Urban Talent of Rotterdam University of Applied Sciences, the Netherlands.

Michael J. Scheel is a Professor and Chair of the Department of Educational Psychology at the University of Nebraska-Lincoln, USA.

Krysia Teodorczuk is a Clinical Psychologist in private practice in Johannesburg. Her areas of interest include mental health with a special focus on adolescent well-being, strength-focused interventions, psychotherapy, and working through trauma.

Dr. Nicole Torka is a policy advisor on "Work, Organisation & Health" at the University of Twente, Enschede, the Netherlands.

The pursuit of happiness*

Hubert Hermans and Frans Meijers

The pursuit of happiness is becoming a central topic in a globalising society in which more and more people seek to fulfil the promises of political freedom and growing financial resources. When material opportunities are available to larger segments of the population, people are not only concerned with the question whether their basic needs are fulfilled, but also, and to an increasing degree, whether they are living a happy life. Indeed, the obsession with how to be happy is reflected by the growing piles of self-help books in bookstores all over the world and in the millions of copies of happiness books that are sold.

The pursuit of happiness is not only of growing concern in the broader population. Psychologists have also been increasingly productive in studying the main features of this phenomenon. Maybe to the surprise of some readers, scientists have collected evidence demonstrating that it is a misunderstanding to think that the pursuit of happiness is simply a bourgeois concern, an expression of self-centeredness, or a factor irrelevant to psychological adaptation and personal development. On the contrary, a growing body of evidence shows that the ability to be happy and contented with life is a central criterion of positive mental health and social adaptation. Reviews on this topic (Diener, Lucas, & Scollon, 2006; Lyubomirsky, King, & Diener, 2005; Lyubomirsky, Sheldon, & Schkade, 2005) suggest that happiness has many positive byproducts that have beneficial consequences for individuals, families, and communities. Happy individuals gain tangible benefits in a variety of life domains. They experience larger social rewards: more friends, stronger social support, and richer social interactions. There are indications that happiness is associated with superior work outcomes: increased productivity and creativity, more activity and flow, higher quality of work, and higher income. Contrary to the idea that happy people are just self-centered or selfish, the literature suggests that happy individuals tend to be relatively more cooperative, prosocial, and charitable. Moreover, subjective happiness may be integral to mental and physical health. Happy people are more likely to evidence greater self-control, have a bolstered immune system and even live a longer life.

Apart from the question of what are the benefits of happiness or subjective wellbeing (the terms are often used interchangeably), there is another challenging question to pose that is at the heart of the present special issue: How can happiness be increased and then sustained? Stimulated by this question, Lyubomirsky, Sheldon, et al. (2005) proposed a partly optimistic model as a response to the prevailing pessimism engendered by the well-known factors of genetic determinism and hedonic adaptation (the observed tendency to quickly return to a relatively stable level of happiness or unhappiness despite major positive or negative events or life changes, a phenomenon also known as the "hedonic treadmill"). Is there some space for increasing one's happiness beyond the stabilising nature of these determinants?

Lyubomirsky, Sheldon, et al. (2005) presented their model under the banner "the architecture of sustainable happiness" (p. 114). They proposed that a person's chronic happiness level is governed by three major factors: a genetically determined level of happiness, happiness-relevant circumstantial factors, and happiness-relevant activities and practices. The basic level is *genetically determined* and is assumed to be fixed, stable over time, and immune to influence or control (unless future scientists will learn how to alter people's basic dispositions and temperaments). *Circumstantial factors* refer

*This is the last article that emerged from the cooperation between Frans Meijers and me (HH). I remember him as an intimate friend, productive colleague, ardent professional, and dedicated scientist. This article and special issue as a whole are published as a tribute to Frans in memory of his precious gifts to the world.

to the individual's personal history, that is, life events that can affect one's happiness, such as a child-hood trauma, an automobile accident, or winning a prestigious award. Circumstantial factors also include life status variables such as health, income, marital status, occupational status, job security, and religious affiliation. The third and arguably most promising means of altering one's happiness level is *activities and practices*. Intentional activities do not happen by themselves but require some degree of effort. One of the critical distinctions between the category of activity and the category of life circumstances is that circumstances happen to people, whereas activities are ways that people act on their circumstances. Some types of behavioural activities, such as exercising regularly or trying to be kind to others, are associated with increased wellbeing. Some types of cognitive activity, such as reframing situations in a more positive light, and particular kinds of volitional activity, such as striving for important personal goals, have the potential of increasing happiness levels.

Taking existing evidence into account, Lyubomirsky, Sheldon, et al. (2005) suggest that genetics account for approximately 50% of the happiness variation, circumstances for approximately 10% and intentional activity for the remaining 40%. They consider these percentages as supporting their proposal that intentional activities offer a possible route to more enduring increases in happiness beyond basic determinants. In other words, changing one's volitional efforts may have a happiness-boosting potential that is almost as large as the probable role of genetics, and apparently much larger than the influence of one's circumstances.

The happiness-increasing effect of intentional activities brings us to the heart of this special issue. As (guest) editors of this journal, that has guidance and counselling as its main purpose, we were interested in theories, methods, and empirical work that would contribute to understanding some basic mechanisms in the increase of happiness and provide practical guidelines to achieving this goal. Therefore, we invited colleagues from very different parts of the world who are doing work that has direct relevance to the third category: intentional activities. We did so in the expectation that the combination of theory, method, and empirical work would be well-suited to offering new insights and practical guidelines for both scientists and practitioners interested in the development of long-term subjective wellbeing.

Most contributions to this issue, but not all of them, are inspired by Dialogical Self Theory (Hermans & Hermans-Konopka, 2010) that conceptualises the self as a dynamic multiplicity of relatively autonomous *I*-positions in the society of mind. This theory is particularly suited to the study of intentional activities as it considers the variety of volitional practices in which a person is engaged as originating from relatively independent intentional *I*-positions. According to this theory, not only the content of the different *I*-positions (e.g. I as a sports fanatic, I as a helpful colleague, I as wanting to become a doctor) is relevant to mental health, but also the way they are *organised* as parts of a dynamic self, including the relative dominance of *I*-positions, neglected *I*-positions, and the role of significant others as internalised *I*-positions in the self. Moreover, this theory has generated a diversity of methods and practical procedures for the stimulation of learning processes in education (Meijers & Hermans, 2018) and for the reorganisation of the self in psychotherapy (Konopka, Hermans, & Goncalves, 2019).

Introduction to the contributions

Katrin den Elzen is interested in the question of how happiness and wellbeing can be achieved when someone faces extreme adversity such as severe disability and multiple loss. She peruses autobiographical material from Nick Vujicic who was born without limbs and from Barbara Pachl-Eberhart who lost her whole family in a car accident. Den Elzen analyses the authors' autobiographical texts from the perspective of Dialogical Self Theory, and portrays how they found inner courage that enabled them to give life-affirming responses to their tragedies.

In his exploration of the renewal of the self as a multi-voiced process, Michiel de Ronde suggests to his clients that they read the New Testament story of the Prodigal Son, and identify themselves with the position of the three protagonists: the youngest son who asks for his heritage and leaves, the happy father who rewards his returning son with a great feast, and the oldest son who becomes discontented as he has devoted his best efforts to his duties during all those years. De Ronde demonstrates how the triple reading of the story of the father with the two sons helps him and his clients to transcend any black-and-white thinking and shows how the dialogue between the different positions is filled with shadows.

Jennifer Hausen starts her contribution by noticing that in contemporary culture it is natural to think that purchasing and owning the "right" possessions results in happiness. In contrast to this expectation, several lines of research demonstrate that high consumption lifestyles and materialistic values are not to be considered trustworthy paths to wellbeing. As a viable alternative to the pursuit of happiness, she proposes Minimalism as a way to find happiness in materially simpler lifestyles. She demonstrates how decision-making processes in the transition from a materialist to a minimalist lifestyle may profit from recognising the self as moving between multiple and relatively autonomous I-positions.

In their contribution, Olga Lehmann, Goran Kardum and Sven Klempe explain the human search for "eudaimonia" as involving the search for inner silence. In doing so, they suggest that the awareness of the dynamics of different I-positions can empower the person to feel more freedom over their thoughts, feelings, and actions. The practice of inner silence can also promote the experience of genuine dialogues with others or with ourselves. The authors describe a number of implications for counselling associated with self-exploration and self-transcendence, such as meditation, contemplation, and prayer.

Working as a psychotherapist, Georgia Gkantona introduces the so-called Dialogical Stage Model, derived from Dialogical Self Theory, as a procedure for evoking positive counter-emotions as responses to prevailing negative emotions, such as sadness, guilt, fear or anger. Dialogical relationships between emotions are facilitated in order to create space in the self for the emergence of positive emotions of happiness, hope, affection, intimacy and love. This method requires the therapist and the client to be fully present in the moment as a comprehensive embodied living person.

Nicole Torka presents the thesis that honesty towards ourselves and others is a pre-condition for genuine happiness. She argues that genuine happiness is impossible without authentic concern for the wellbeing of others. She considers honesty as a distinctive and universal virtue that not only contributes to the authentic fulfilment of self-motives and other-motives, but also to the common good. Such an incorporation of others into the self underscores a "democratic self" as adding value to the common good. She holds that the honesty of professionals who work in an educational or vocational setting is vital not only for the good of the self and the other but also for the common good.

As a starting point of their empirical research, Małgorzata Łysiak and Małgorzata Puchalska-Wasyl advocate the idea that the self has the ability to imaginatively move to a future point in time and then speak to oneself from there about the sense of what one is doing in one's present situation. On the basis of this proposition, they invite their respondents to become engaged in dialogues between their future and present I-positions and between their past and present I-positions. They then present data demonstrating that such dialogues have particular adaptive functions, such as providing supportive messages to oneself, redefining the past, and taking distance from current experiences.

Evelyn Plumb, Kathryn Hawley, Margaret Boyer, Michael Scheel and Collie Conoley's article is focused on empirical support for Goal Focused Positive Psychotherapy (GFPP), a comprehensive, evidence-based, psychotherapy model that is inspired by positive and social psychology research. The approach puts special emphasis on idiosyncratic and multiculturally-attuned client factors, particularly client strengths and goals, in order to increase subjective wellbeing and facilitate the client's experience of a meaningful, satisfying life. Instead of treating client deficits, symptoms, or trauma, the method focuses on procedures for enhancing client wellbeing.

Another contribution from positive psychology is Krysia Teodorczuk, Tharina Guse, and Graham du Plessis's study on the effect of interventions on hope and wellbeing among adolescents living in a child and youth care centre in South Africa. The experimental group partook in one-hour intervention sessions weekly, for a six-week period. No statistically significant differences in wellbeing and hope were found between the experimental and control group. The authors discuss moderating factors and offer reflections to better understand these outcomes.

Dave Redekopp and Michael Huston are interested in work as a significant factor in mental health and wellbeing outcomes. They argue that career development processes can be helpful in finding and managing work trajectories that lead to these outcomes. They review evidence for the interactive relationships between work, career development, mental health and mental illness. They also provide evidence for counselling and guidance intervention, organisational changes and policy directions and conclude with suggestions for improving wellbeing via career development procedures.

In the final part of this special issue, Anuradha Bakshi interviews Ed Diener, one of world's most renowned investigators of happiness and wellbeing who has devoted most of his professional career to unveiling much of the complexities of this phenomenon that is increasingly significant to the live trajectories of individuals in their search for direction and meaning in life. One of the surprising findings mentioned in this interview is that happiness and unhappiness are not to be understood as opposites but as relatively independent dimensions.

Lastly, Anjali Majumdar, Satishchandra Kumar and Anuradha Bakshi review Martin Seligman's autobiography, the Hope Circuit. The founder of Positive Psychology, Seligman, charts his own life in the making in concert with psychology in the making, particularly with regard to the paradigm shift from deficit to asset models.

Disclosure statement

No potential conflict of interest was reported by the authors.

References

Diener, E., Lucas, R. E., & Scollon, C. N. (2006). Beyond the hedonic treadmill: Revising the adaptation theory of well-being. *American Psychologist, 61*, 305–314.

Hermans, H. J. M., & Hermans-Konopka, A. (2010). *Dialogical self theory: Positioning and counter-positioning in a globalizing society*. Cambridge, UK: Cambridge University Press.

Konopka, A., Hermans, H. J. M., & Goncalves, M. M. (2019). *Handbook of dialogical self theory and psychotherapy: Bridging psychotherapeutic and cultural traditions*. New York: Routledge.

Lyubomirsky, S., King, L., & Diener, E. (2005). The benefits of frequent positive affect: Does happiness lead to success? *Psychological Bulletin, 131*, 803–855.

Lyubomirsky, S., Sheldon, K. M., & Schkade, D. (2005). Pursuing happiness: The architecture of sustainable change. *Review of General Psychology, 9*, 111–131.

Meijers, F., & Hermans, H. J. M. (2018). *The dialogical self theory in education: A multicultural perspective*. New York: Springer.

Finding happiness and wellbeing in the face of extreme adversity

Katrin Den Elzen

ABSTRACT

This paper examines how people find happiness and create wellbeing when confronted by extreme adversity. Utilising the Dialogical Self Theory for an analysis of published autobiographies, it investigates two case studies, Nick Vujicic, who was born without limbs, and Austrian author Barbara Pachl-Eberhart, who narrates creating a fulfilled life after losing her whole family, her husband and two young children. Drawing on the DST process of positioning and repositioning to renew the self, this research investigates positions that allow the authors to work through their adversity to establish happiness and wellbeing. The aim of this study is to contribute knowledge to research on individual well-being, in particular following adversity.

Introduction

Happiness and wellbeing are goals that are particularly difficult to achieve when someone faces extreme adversity such as severe disability and multiple loss. This paper researches two case studies, Nick Vujicic and Barbara Pachl-Eberhart, and applies the Dialogical Self Theory (DST) to an analysis of the author's autobiographical texts in order to examine how people find happiness and create wellbeing when confronted with serious adversity. The two autobiographies by Vujicic, *Life Without Limits* (2011) and *Unstoppable* (2012), portray how the author, who was born without limbs, overcame enormous challenges in dealing with his disability to build a life that he describes as "ridiculously happy" (2011, p. viii). Austrian author Pachl-Eberhart's memoir (2010) *vier minus drei* (four minus three)[1] as well as her second book *Warum gerade du? (Why you?)* (2014), depict her journey of recovery after losing her whole family in a car accident and are noteworthy for Pachl-Eberhart's inner courage and life-affirming response to the tragedy. In addition to the books, I will include interviews with Vujicic and Pachl-Eberhart in my research.

Key theorist in the field of narrative psychology Hubert Hermans developed DST, which frames the dialogical self as a "dynamic multiplicity of I-positions" that can be compared to interpersonal dialogue. Subject positions are seen to be involved in dialogue and negotiation (Hermans & Gieser, 2012, p. 2). Central to DST is the notion of voice and "multivoicedness", which conceptualises voice as "the tool by which the necessary relationship of communication is established" (Salgado & Hermans, 2005, p. 11). Positions are not fixed but are constructed depending on context and its present situation in time and space. "Positions are not to be understood as stable centres of knowledge, but as perspectives that vary with the direct interchange with the social environment" (Hermans, 2008, p. 193). This research will identify the autobiographical representation, juxtaposing and negotiation of subject positions that create happiness and wellbeing. At the heart of this investigation is a particular focus on the positioning of identities that have arisen out of extreme adversity and how they are repositioned to build well-being and inner happiness. Utilising DST for an examination of literary texts rather than clinical practice constitutes a new approach. This investigation will show that the two authors find happiness under extreme life circumstances. It suggests that the narrative process of innovation and dialogical interchange is capable of transforming the grieving self to create wellbeing and enduring happiness.

The aim of this study is to contribute knowledge to research on individual well-being, in particular following adversity. With this focus in mind, this paper will draw on the philosophical concept of the "art-of-living" in order to theorise happiness further. Philosophy and life writing offer "a fertile conjunction" that generates possibilities of study by addressing the "often elusive affiliations between the theoretical and the personal, the abstract and the concrete" (LeMahieu & Cowley, 2018, p. 301). Art-of-living is an old concept that dates back to classical antiquity and that has been revived by modern philosophers, in particular the German philosopher Wilhelm Schmid (2008, p. 209). Schmid conceptualises the art-of-living as a way of living a good, meaningful life that requires the individual to think and reflect about their life and to act and choose in a deliberate and conscious manner (2008). Schmid positions life as art, because he sees it as a creative activity and an ongoing process of learning. He states that there is no system that can provide objective answers, but that philosophy can offer guidance in realising this conscious lifestyle (2008, p. 209).

Born without limbs

This section focuses on investigating the process of innovation of the dialogical self that is represented in Vujicic's written and video material. Innovation is a core concept of DST, which contends that the dialogical self is able to renew itself. The process of positioning and repositioning enables innovation. Identity construction is conceptualised as the result of dialogical interchange that makes it possible for new perspectives, contexts and understandings to arise, thus leading to the re-evalutation, reconstruction and reconciling of existing positions (Hermans, 2008, pp. 192–194). Vujicic's two autobiographies narrate how he created a "Life without Limits" despite the enormous challenges inherent in having no limbs and having to cope "in a world made for arms and legs" (2011, p. 19). He positions his life as "ridiculously good" and explains that he experiences joy every day irrespective if things go smoothly or not, even if he is feeling sick (2012, p. 123). The texts clearly and explicitly narrate the representation and negotiation of Vujicic's subjectivity, the rebuilding of identity continuity out of a self that was fragmented by disability, and are well-suited to an analysis of the process of identity construction.

The disabled child: from suicidal thoughts and fear to self-acceptance

Vujicic begins his narrative with a representation of his birth and childhood and portrays himself as being born without limbs in December 1982 (2011, p. 4). He positions himself as happy and carefree when he was so young that he was unaware of his disability and differences. This shows that a position is located in a specific time and depends upon the context, represented here as the absence of an awareness of being different. *Life Without Limits* includes colour photographs of Vujicic at different ages. One of them depicts him as a smiling, happy six months old. The caption reads "Happy, confident and cute – right? My blissful ignorance was a blessing at that age, not knowing that I was different or that many challenges awaited me" (2011). However, once Vujicic was old enough to understand that he was different and in particular had become exposed to the negative reaction of others, his perception of selfhood became negative. Other children evoked insecurities by asking him why he did not have arms or legs (2011, p. 17) and assumed that he was mentally disabled (2011, p. 18). DST conceptualises identity construction as "a relational or dialogical production" and sees the self emerging in reference to and mutual dependency of others (Salgado & Hermans, 2005, p. 11). Vujicic's position of young child, which did not perceive himself as disabled, undergoes a change in perspective as a result of his interaction with external others who bring his disability into sharp focus for him and lower his self-worth.

In addition to the physical challenges inherent in having no limbs, Vujicic's disabled position had to deal with many emotional challenges during his childhood, such as a range of fears: the fear of rejection, of inadequacy, and above all of being dependent (2011, p. 119). One of Vujicic's greatest fears growing up was being a burden on the people he loved, in particular, his parents and siblings: "I was nearly overwhelmed by visions of dependency" (2011, p. 120). He repeatedly represents his position of disabled child as being haunted by an existential worry of being a burden throughout his texts: "I was absolutely convinced that my life would never be of value and that I would only be a burden to those I loved" (2011, p. 39). In contrast to the awareness of his disability by the young child, which was prompted by the interaction with others, the disabled position's fear of dependency is not a result of external others, but of intrapersonal communication prompted by an awareness of his real physical limitations.

According to Hermans, the dialogical self is "continually challenged or plagued by questions, disagreements, confrontations and conflicts because other people are represented in the self in the form of voiced positions functioning as centres of initiative, construction and reconstruction" (2008, p. 193). The disabled self searched for meaning and is positioned as questioning God: "Why couldn't You give me just one arm?" (2011, p. 16). The disabled identity is juxtaposed in opposition to his religious position: "I felt I was just a mistake, a freak of nature, God's forgotten child" (2011, p. 48). He positions himself as alienated from and challenging God: *Did I do something wrong? … Why won't you help me? Why do you make me suffer?*" (2011, p. 46, emphasis in original). In his adult life, Vujicic's religious identity becomes a dominant position, but growing up, it is represented as confused and unsure. Vujicic narrates that he received a religious upbringing by his parents, Serbian migrants to Australia who came from strong Christian families (2011, p. 9). His religious identity was instilled by his parents in a loving way, who believed "that God had a plan for him and one day He would reveal it" (2011, p. 6). The religious identity of the young Vujicic experienced contradictory viewpoints, whereby his own perception of being God's mistake is placed in conflict with his parents' view of a loving and purposeful God, whose voiced positions function as a centre of initiative in their son's dialogical self. DST suggests that the self is constructed out of different positions who constitute different worldviews with their own desires and anxieties (Hermans, 2001, p. 332). The disabled position's fear of dependency and questioning God expresses opposite emotions and perspectives to the loving parents who are framed as supportive and trusting of God.

Vujicic positions his strong fear of being a burden on his family as the impetus behind his suicide attempt at ten years of age (2011, p. 49). In addition to perceiving himself to be "a tremendous burden" (2011, p. 48), the disabled child is shown to be haunted by existential questions, in particular his worry that he might never be able to find a wife, asking himself: "Would any girl want a boyfriend who could not hold her hand or dance with her"? (2011, pp. 44–45). The lack of explanation, in particular, a scientific one, for his disability added to his feeling of hopelessness (2011, p. 46). Negative thoughts became more and more common: "I was haunted by dark thoughts that if I could not change my body, I'd end my life" (2011, pp. 147–148). The suicide attempt by the disabled child is narrated in detail and positioned as a pivotal experience and turning point in his life. He tried to kill himself by submerging his head under water whilst taking a bath, but in that crucial moment, he realised that he could not bear to cause his loving family pain (2011, p. 50). The suicidal disabled self is thus repositioned and re-evaluates his perspective, putting love above his suffering. Hermans frames spatiality as a core feature of the dialogical self, and selves are conceptualised by the processes of "positioning" and "repositioning" (2001, p. 329). Central to the spatiality of positioning is the notion of "emplotment", which makes the juxtaposing of events possible" (Hermans, 2001, p. 341).

The process of repositioning that the suicidal disabled child undergoes is shown to lead to a renewal of the disabled self, who only two years after his suicide attempt changed his outlook on his life. Innovation of the self is created when the disabled self repositions his interpretation of his disability from his physical limitations to his capabilities: "I had learned to focus on my abilities instead of my disabilities" (2011, p. 145). This is a significant and pivotal re-evaluation of his disability.

According to DST, a core feature of spatial positioning is that selves evaluate the events that they portray and juxtapose. The mere act of identifying positions does not reveal "which stories are told and which meanings are expressed from the perspective of" that particular voice (Hermans, 2001, p. 335). It is necessary then to analyse the meanings that are assigned by a particular voice in order to understand it.

The disabled self has undergone a repositioning of his fear of dependency and positions his recovery from his suicidal despair as a result of firstly his acceptance of his disability and secondly of the role his loving family played: "I'd accepted my lack of limbs and I'd managed to become a pretty happy and self-sufficient kid" (2011, p. 145). Schmid argues that acceptance of life's conditions in all its polarity is a vital aspect of the art-of-living. He distinguishes between what he has termed "feel-good-happiness", which he views as feeling content, happy and healthy, and the "happiness of plentitude" (2007, pp. 18+29). By this he refers to the awareness that life is polar, that it takes place in opposites, that there is not only success but failure, not only joy but pain. He frames the deciding question: Can I accept that? If a person can accept polarity, then this makes the happiness of plentitude possible, which Schmid (2015) conceptualises as the only type of happiness that is enduring. Fulfilment does not come through happiness alone, but also by being willing to experience unhappiness. Schmid advocates that we have to be able to be unhappy, that it is a normal part of life. He also contents that unhappiness can be a catalyst for new perspectives and ideas (2015), which mirrors Vujicic's experiences.

Vujicic assigns his parents an instrumental voice in dealing with his disabilities in a positive way, who "encouraged him to acknowledge his fears and then push past them" (2011, p. 120). He credits his family's love and support as having been constitutive of his happiness: "I'm blessed in more ways than I can count. I think life would be far more difficult for someone who lacks a loving family like the one I had" (2012, p. 89) and portrays his parents as "his pillars of strength" (2011, n.p.). The DST concept of "the-other-in-the-self" frames the other not as external and "added" to the self, but rather as constitutive of the self (Hermans, 2008, p. 186). In addition to his parents, Vujicic credits his siblings and close-knit extended family as contributing to his childhood as a happy one, positioning their love as a gift: "what a powerful gift it is to be loved like that" (2011, p. 16), thus framing them as constitutive of his selfhood.

Creating happiness: the motivational speaker and evangelist subject positions

Vujicic positions two dominant identities as instrumental to his happiness and wellbeing: International motivational speaker and evangelist with a global ministry. DST frames positions as being "involved in relationships of relative dominance and social power", which results in some positions being dominant and others being repressed (Hermans & Gieser, 2012, p. 2). Although these positions are closely related, they are also distinct, in that they have diverse goals and somewhat different worldviews. The evangelist has a clear intent to preach the love of God, and Vujicic's non-for-profit organisation "Life Without Limbs" highlights on its website that the organisation only accepts Evangelical Christian faith-based speaking requests (2018b). Even though the motivational speaker talks about God, the main focus in his speeches and books is on inspiring his audiences and to offer them a clear methodology to achieve a happy and fulfilling life. Vujicic has created two different organisations with their own websites that fulfil the mission of each position. He founded the company "Attitude Is Altitude" to handle his corporate speaking engagements, and the name is intended to denote the importance of a positive attitude (2011, p. 91). The website describes Vujicic as follows:

> Nick is an internationally renowned motivational speaker who travels the world sharing his message of hope. Nick has visited over 65 countries, met with 16 presidents, addressed 9 governments and spoken live to over 6.5 million people. He is an accomplished author with five titles published in over 30 languages including the New York Times Best Seller "Unstoppable". Nick's unique story and enthusiasm for life will inspire you towards positive change! (2018a)

According to DST, positions can be juxtaposed in various ways, including in conflict and in agreement (Hermans & Gieser, 2012, p. 2). The two identities of speaker and evangelist are positioned in agreement in that they both pursue the same goal of contributing to positive change in the world, and as such are mutually supportive of one another. The target audiences of each are different, one aims at Christians, and the other at corporate, private, government and educational sectors.

The despairing child has undergone a process of innovation and renewal to become a successful motivational speaker and evangelist addressing millions of people globally: a search on YouTube with the term "Nick Vujicic" yields 399,000 results (20 April 2018), which indicates how well-known Vujicic has become. As a sixteen-year-old, Vujicic was invited to speak at a Christian youth discussion group, where he shared his earlier belief that he was one of God's rare mistakes, but that he was slowly learning that this was not true (2011, p. 88). When Vujicic was startled by the reaction of his audience, who nearly all cried, the group leader explained that he was that good a speaker that he moved his audience. What followed from this initial talk was dozens of invitations to speak to church groups and youth organisations. As a result of these invitations and the emotional reaction by his audience, Vujicic realised that he had "riches to share, blessings to lighten the burden of others" (2011, p. 89). The early motivational speaker position re-evaluated his selfhood and repositioned his earlier fears of dependency and limited life opportunities into a realisation of his self-worth and personal talents. This significant re-evaluation is characteristic of the spatial nature of the dialogical self. An essential feature of spatial positioning is the evaluation by the narrator of the events that are portrayed and juxtaposed (Hermans, 2001, 341). A position's interpretation refers "to the active processing of giving positive or negative value to the events in one's life" (Hermans, 2001, p. 335). Vujicic positions his early attempts at gaining paid speaking engagements as difficult, in that he was often turned down as being "too young or too inexperienced or just too unusual" (2011, p. 130). Through reflection, the budding speaker evaluated that he had discovered his calling: "Every time I was turned down it hurt so much I realized that I'd found my passion" (2011, p. 131).

Living his passion, which is framed as fulfilling his calling of helping others, is portrayed in his books as a position that is the catalyst that drives the speaker position: a whirlwind career as an international speaker that sees him addressing "tens and hundreds of thousands" of people (2011, p. 118). At only 20 years of age, he embarked on an international speaking tour in South Africa. Vujicic represents this tour as a pivotal time in his life: "It helped me realize what I wanted to do with the rest of my life: to share my message of encouragement and faith around the globe" (2011, p. 218). He positions encountering enormous poverty, "children with open wounds from flesh-eating bacteria" and people dying of AIDS, as such an eye-opening experience that human suffering leads him to re-evaluate his own situation: "It was so much worse than anything I've ever endured, and it made my life seem pampered by comparison" (2011, p. 219). Hermans argues that identity negotiation takes place through the process of innovation. One of the ways in which innovation can be achieved is through the introduction of new positions, both internal and external (2004, p. 178). The experience of the trip evoked a major re-interpretation of Vujicic's perception of his own suffering, which leads him to reflect that he led a very "self-centred and selfish existence" (2011, p. 219). He attributes South Africa as changing him, recognising that he was so pre-occupied with his disability "that growing up he could not conceive of anyone suffering as much as himself" (2011, p. 220). His self is renewed through the process of innovation, whereby the poverty and suffering of others are introduced as positions into his dialogical self, causing him to instrumentally reorganise his own disabled position into gratitude for all that he has in his life, in particular, the love of his family. This repositions his earlier perception of his victimhood into a new interpretation of self-centredness. In addition, the external others who experience poverty and suffering evoke an additional innovation of Vujicic's selfhood: his compassionate self. He expresses his amazement that on his travels, people "in stark poverty and great suffering" react with incredible compassion and empathy towards him (2011, p. 105). While he travelled to South Africa as a motivational speaker, he attributes the people with teaching him compassion, by being empathetic towards him despite their own hardship. This leads to Vujicic's realisation that he naturally has a connection with others through his disability that bridges cultural and linguistic barriers.

As a child Vujicic's identity had been fragmented by fears of dependency and being a burden to others. As an adult this position has undergone such striking re-evaluation that the disabled identity is now positioned as the very impetus for his happiness. This position has re-evaluated the perception and self-worth of being disabled and a burden to the point where the new worldview, expressed through the eyes of the motivational speaker, actually has an opposite interpretation: "I'm officially *disabled,* but I'm truly *enabled* because of my lack of limbs" (2011, p. 2, emphasis in original). He explains this interpretation of his disability further: "I can't even say anymore that my ridiculously good life has come about despite my disabilities and the hardships I've faced. Now, I must say that my grand life is because of my disabilities and hardships" (2012, p. 66). By that he means that his uniqueness has enabled him to connect with others because they only have "to look at me to know I'd faced and overcome my challenges" and this gave him instantaneous credibility (2011, p. 21). Vujicic frames his challenges as instrumental to his inner happiness: "I honestly appreciate life more because I've had to struggle to do many things that most people simply take for granted" (2012, p. 66). His attitude is characteristic of the art-of-living, which suggests that an openness to life which embraces polarity and adversity leads to enduring happiness, which differs from the superficial feel-good happiness that cannot last (Schmid, 2015). Vujicic adds that there are many "people with perfect bodies who don't have half the happiness I've found" (2011, p. 29). While others who bullied him were positioned as detrimental to his wellbeing as a child, now a re-evaluation has taken place that sees the disabled self build his self-worth on the positive reaction of others. This re-evaluation and repositioning prompted by the new context of motivational speaker who passionately helps others placed the disabled self into agreement with his self-worth. Vujicic's reinterpretation of his disabilities from childhood despair to catalyst of enduring happiness shows the ability of the dialogical self to innovate itself and to rebuild identity continuity. Thus, evaluating the self and events spatially in a way that places adversarial positions into agreement can potentially lead to happiness.

Creating happiness through helping others: the author of motivational books

Another self that is positioned in alliance with the motivational speaker self is Vujicic's author identity. The author has deliberately chosen a structure with his two autobiographies that facilitate his intention to offer the reader a methodology for finding their "*own* purpose and pathway to a ridiculously good life" (2011, p.vii, emphasis in original). His books can be said to be hybrid texts that combine autobiography with self-help. The hybrid structure is well-suited to his dominant speaker position of fulfilling his life purpose. Both the author and speaker positions hold the worldview that helping others is instrumental to happiness: "The greatest rewards come when you give of yourself. It's about bettering the lives of others, being part of something bigger than yourself, and making a positive difference" (2011, p. 27). Schmid conceptualises the act of caring for others as an act of self-fulfilment and contends that inner richness comes as a result of helping others, beyond immediate self-interest (2008, p. 218).

The structure of the autobiographies is such that Vujicic narrates his personal experience of overcoming challenges and then turns to address the reader directly using second person to evoke reflection and to offer concrete self-help strategies. For example, after positioning the relationship with his parents as constitutive of his happiness and quality of life, he prompts the reader: "Take time now to evaluate your people skills, the quality of your relationships, what you put into them … . Are you putting into relationships as much as you take out?" (2011, p. 179). The hybrid structure chosen by Vujicic is juxtaposed in alignment with fulfilling his purpose of helping others. The many YouTube clips about Vujicic are a testament that he has accomplished his quest to lead a life that is made meaningful and fulfilling by helping others. One of the most popular videos on YouTube (published 12 March 2010) had over 65 million views by April 2018.[2] In addition, Vujicic narrates that he receives many emails from his viewers and readers who attribute his motivational messages to improving their lives significantly (2011, p. 28).

Whilst the motivational speaker provides his audiences with inspiration and practical self-help strategies to achieve happiness, the evangelist position desires to inspire others to find God, stating on his website: "If just one more person finds eternal life in Jesus Christ … it is all worth it" and notes that he has brought the word of God to 60 million people (2018b). Like Vujicic's disabled self, his religious position has undergone a major re-evaluation. As a child he perceived himself as a "mistake, a freak of nature, God's forgotten child" (2011, p. 48). The adult religious position no longer conceptualises himself to be forgotten: [God] "never leaves me. He hasn't forgotten me. He will cause even the worst things to come together for the good. … I know that God is good" (2011, pp. 44–45). The innovation of the dialogical self can take place through a "change in dominance relations", whereby the domination of one position over another is mediated, which allows the suppressed position to move to the foreground (Hermans, 2004, p. 178). Vujicic's religious position has undergone a change in power relations. During childhood, it had been suppressed by the disabled position. In adulthood, the religious self has become a dominant position that is no longer in an adversarial relationship but in a coalition with the disabled self, thus denoting the innovation of Vujicic's selfhood through a change in power relations.

Creating happiness through love

There is another dominant position that Vujicic represents as instrumental to his happiness, and that is love. Vujicic positions his love for his wife and sons as integral to his happiness. He describes his wife as "my greatest gift and joy ever" and as "the love of my life" (2012), thus positioning her as constitutive of his subjectivity. Vujicic's third book, *Love Without Limits* (2014) is co-written with his wife and continues his motto of living, and in this case loving, "without limits". The couple have two sons, and Vujicic portrays their family life as "more wonderful then I'd ever dreamed possible" (2014, p. 7). Through the external other, his wife, the position that feared it would miss out on being married is transformed into deep inner fulfilment and happiness. The loving husband and father position is juxtaposed both in alliance and in conflict with his motivational speaker and evangelist positions. It is in alliance, because it was in Vujicic's capacity as speaker that they met. In addition, Vujicic frames the love between two people as God's "gift of lasting love" (2014, p. 3). It is conflictual because it is difficult for Vujicic to juggle his busy schedule as an international speaker with sufficient family time. He conveys that on a "rewarding but gruelling four-month tour of twenty-six countries" he missed his family to the point where it became unbearable: "There are no words to describe how difficult it was to be away from my wife and [his son] Kiyoshi that long" (2014, p. 7). This prompts him to seek more balance between work and family time. As subject positions express context-dependent understandings, the motivational self that desires to address worldwide audiences is reconstructed under the new circumstances of marriage and fatherhood. Vujicic's dominant positions of motivational speaker, evangelist and loving family man together create his "ridiculously good life", whereby he feels "blessed with joy and love beyond measure" (2012, p. 3). Schmid foregrounds the importance of making connections with other people as the path to creating a meaningful life. Such a connection is felt the strongest between people who love one another (2007, pp. 52–53).

Four minus three: rewriting multiple loss

Pachl-Eberhart's book *four minus three* is a grief memoir that narrates the loss of her whole family at 33, when her husband and their two young children aged six and two die in a car accident and focuses its lens on finding happiness again after multiple loss (2010). Pachl-Eberhart narrates the positioning and repositioning of her post-loss identities and feelings in poignant honesty. This section examines her journey of recovery that allowed the self that survived multiple loss to find enduring happiness, an experience of such adversity that she was often told by others: "You experienced the worst that can happen to a human being". Her reply frames this sentiment as too simplistic: "No, I thought time and again, you cannot say that just like that" (2010, 196).

Taking care of the self

Pachl-Eberhart's bereaved self is positioned as being capable of communicating her needs and opening up to others. Hermans conceptualises autobiographical narrators as "active processors of experience" (2004, p. 177). Even in the early stages of grief, Pachl-Eberhart shows herself to have agency and acts in accord with her own wellbeing. The bereaved position takes the active step of writing a long email to everyone listed on her computer, both friends and associates, on 25 March 2008, just five days after the accident and one day after her son's death, entitled *RE: death and its transgression* (2010, pp. 45–56). She describes what happened over the Easter weekend, on Thursday, when the accident happened, on Sunday, when her daughter died in hospital, and Monday when the machines were turned off that kept her son alive. She chooses neutral, unsentimental positioning: "I met death three times in the last five days" and deliberately informs everyone, "I wasn't there … I'm alive and want to remain part of 'normal life'. Please do not be afraid to confront me with this life. … it feels good to feel life in all its facets!!!!" (p. 46). The grieving self takes this active step because she is worried about being ostracised because of the magnitude of her loss:

> If there still was one remaining concern in me, who had nothing left to loose, it was the fear of being speechless. A fear that friends would avoid contact with me because they did not know how to react to me. … that I could become an outsider. (2010, p. 57)

Positioning death as rendering the bereaved mute (2014, p. 12), her motivation for writing the email is framed to be a conscious act of taking care of herself: "I did not want to be left speechless, wanted to counteract speechlessness, so others don't have a psychological barrier. I wanted to explain what happened, what I needed, and what went on in my head" (Judi1783, 2013). Pachl-Eberhart's spatial positioning of the bereaved position with the self that communicates her needs to others indicates that the juxtaposing of an adverse position in agreement with taking care of oneself makes wellbeing more likely. The concept of caring for oneself was coined by Michael Foucault, and it refers to the capacity of the individual to build "a complex relationship with oneself" (Dohmen, 2003, p. 360). It has become an integral part of the art-of-living. Schmid argues that the ability to care for others depends on the ability to care for the self (2008, p. 218).

Conscious and life-affirming grieving

The introduction of positive messages expressed by external others to the bereaved self is positioned as the catalyst for the birth of a new post-loss identity: the inspiring self that comforts others in their grief. Pachl-Eberhart represents herself as astonished at the heartfelt reaction of others to her email. She is inundated with replies by others who expressed gratitude that she is so open about her grief, stating that they feel comforted by her words and urging her to keep writing about her individual path (2010, p. 73). Pachl-Eberhart's self who recognises that she is in a position to help others through their grief leads to the innovation of her identity. Over time this new self, which, like Vujicic, finds meaning in helping others, becomes a grief advisor, who teaches writing to heal, in seminars and online, and inspirational speaker. Pachl-Eberhart became well-known in Austria, because of the magnitude of her loss, but also because of her individual, life-affirming approach to grieving. She was and continues to be invited to many TV, radio and print media interviews.

Her bereaved position can be characterised as life affirming, with a positive attitude towards life, filled with gratitude for her family: "I live with deep gratitude that three angels decided to accompany me seven years of my life" (2010, p. 54). Even under the context of sudden and multiple loss this position focuses on the love for her family, rather than what was taken from her, thus evaluating her grief in terms of love, which is framed as continuing after death, rather than her loss. The positioning of the bereaved self that mourns premature loss in agreement with gratitude rather than evaluating herself as a victim shows that the process of positioning bereavement in agreement with positive emotion can be supportive of wellbeing.

Importantly, this alliance of the bereaved self with gratitude does not take place in denial of the intense pain evoked by grief. On the contrary, the grieving self is positioned to have been willing to feel the pain in its full intensity. Pachl-Eberhart narrates that she lived initially in a kind of dream-state, which she describes as a bubble of love, where she felt close to her family, and they visited her at night in happy dreams (2010, p. 130). This protected state came to an end, and then the pain is positioned to have arrived without warning: "I know the pain. The inconceivable, unimaginable, deep pain, which threatens to break our heart and to destroy life. I know it well" (2010, p. 156) and the grieving self positions herself as disabled and amputated (2010, p. 271), thus denoting a fragmentation of the self that happened not immediately after her loss, but later on. In her grief, she wailed, shivered, screamed, bit into her pillow or into her knuckles and somehow hung on in there until the attacks calmed down (2014, p. 141). She frames feeling such pain as a lonely journey. Careful consideration through intrapersonal dialogue leads the bereaved self to realise that it was good that the pain demanded of her to be alone: "Because, no matter how many knifes it stabbed into my heart, no matter how much it hurt me", it also brought a present in the form of an insight. "A different perspective. A signpost, which pointed me into a new direction" (2010, p. 157). In her grief, Pachl-Eberhart is able to meet the pain rather than suppressing or resisting it (2014, p. 159). Schmid contends that the ongoing act of composing a meaningful life and self includes being able to be present with and to accept emotions such as trauma, hurt, fear and adversity (2008, p. 213). He goes one step further and advocates befriending one's moods and emotions. According to Schmid, befriending the self means to successfully integrate opposing selves, in order to deal with the striving for dominant power by some selves. Trying to escape the self instead of being present makes befriending the self impossible (2008, p. 215). Pachl-Eberhart's ability to be present with, to accept and even befriend her grief, one of the most intense and distressing emotions, shows her personal path of the art-of-living.

According to DST, remembered others can constitute identities in our subjectivity (Hermans & Gieser, 2012, p. 17). Even in her initial grief, her family continues to be constitutive of Pachl-Eberhart's selfhood through their love as remembered others. This life-affirming attitude is not juxtaposed in conflict with her pain of loss, but embraces it and sees death and living life fully not as mutually exclusive, but rather as belonging together (Judi1783, 2013). Schmid frames the ability to embrace life, in all its polarity, as a characteristic of the art-of-living, including the acceptance of the shadow sides of life (2007, p. 29). In Pachl-Eberhart's consciously chosen open attitude towards multiple loss, which affirms life and its shadow death, we see the theoretical embracing of adversity espoused by Schmid practiced in her life.

Happiness through a meaningful purpose: helping others to navigate grief

In 2014, Pachl-Eberhart published a second autobiographical book entitled *Warum gerade du?* (*Why you?*), which she wrote to help others to contemplate death in a life-affirming way. Whereas her memoir was written in the immediacy of her loss and narrates the first year of her bereavement, *Why you?* was published six years after the death of her family and as such is based on hindsight and her personal insights into grief. In particular it aims to answer the question of how to be happy again following loss. In a structure that is reminiscent of Vujicic's hybrid form of autobiographical writing and self-help, the grief advisor position draws on her own personal bereavement to help others through their grief, thus forming an alliance with the bereaved self. Like Vujicic, she shares her personal experiences of adversity with the reader and the insights that she gained. Autobiographical writing is an act of remembrance and reflection. According to Schmid, the "happiness of plentitude" spans over time and is enduring. As such, it is not a "spectacular" happiness but grounded in a consciousness that is reflective, which is often not accessible in the moment. This type of happiness generally takes place through remembering, from a distance, which creates coherence in one's life, with all its "lightness and shadows" (2007, p. 34). Schmid sees "narrative connections" as particularly suited for meaning-making, in that personal narrative forges connections between events and experiences, often through metaphors (2007, p. 64).

Pachl-Eberhart prompts the reader to engage with existential questions that the bereaved face such as "How can I bear this pain? Why did you have to die? Can I be happy again?" (2014, p. 13). She describes that she learned to bear these questions, to live with them as if they were guests in her home, by avoiding desiring instant answers, and positions this attitude as allowing her to discover new and surprising answers over time. Hermans and Gieser suggest that the dialogical self is constructed out of dialogical relations that engage in interpersonal and intrapersonal dialogue between internal, internalised and external positions (2012, pp. 6–7). Pachl-Eberhart frames her intrapersonal dialogue as being instrumental in the innovation of her selfhood. She frames her insights as small and fragile, but nevertheless sustainable, which "allowed the foundations of her life to be more solid than ever before" (2014, p. 14). The notion of life being "more solid than ever before" indicates the renewal of Pachl-Eberhart's post-loss identity. Further, she represents being engaged with every-day life and others as the catalyst for finding existential answers rather than brooding: "It was in the interaction with the now that I found the meaning in the events of the past. I lived, and I gave myself time" (2014, p. 14). As such, she positions both interpersonal engagement with friends, family and strangers and intrapersonal dialogue as constitutive of her wellbeing following loss.

In *Why me*, Pachl-Eberhart positions her post-loss self as unreservedly happy again (2014, p. 222). The book narrates the process of positioning and repositioning of her bereaved self and the self that believed in and consciously created happiness. She evaluates happiness as small moments of happiness and big happiness, which she interprets as lasting inner happiness. The small moments are positioned as a gateway to the big happiness over a period of time. Pachl-Eberhart juxtaposes being able to open up to brief moments of happiness in alignment with the bereaved self in the initial emotionally raw and intensely painful stages of her grief:

> I was modest, even regarding happiness. I began to collect moments, where I felt just a little bit better than the preceding moment. It was surprising, how quickly I found something. I noticed the warm breeze, the smell of fresh bread. (2010, p. 240)

The bereaved self begins to appreciate and remind herself that it is the little things in life that make up the meaningfulness of life. "A smile. A kind word. An act of kindness" and she frames her grief as beginning to transform: "And it works" (2010, p. 279). This transformation is represented to lead to lasting happiness, which Pachl-Eberhart evaluates as coming from within as a result of having learned to love herself. "Love that we carry in our hearts is part of this happiness, for our deceased loved ones, ourselves, and life itself" (2014, p. 245). In addition to consciously focusing on moments of "small happiness" to provide herself comfort and relief, the bereaved self turned to books on near-death experiences, bereavement and parapsychology in order to make sense of her loss and her spiritual experiences (2010, p. 60). DST suggests that books can function as a catalyst for facilitating understandings in positions (Hermans & Gieser, 2012, p. 17). The insights gained from these nonfiction books function as new positions that are introduced to the bereaved self, thus reorganising that identity, which feels reassured by the information and less isolated in her loss. These deliberate acts by the bereaved self of processing her experiences of loss by juxtaposing it with internal and external positions that are supportive of alleviating her grief indicates that placing the bereaved self in a positive relationship with other positions potentially is constitutive of rebuilding happiness and wellbeing.

The grieving self is positioned to be guided by an inner voice that supports Pachl-Eberhart's wellbeing from the moment she found out about the accident, which is depicted in italics throughout the memoir (2010, p. 70). "The inner voice in my head kept talking to me and was determined to pursue the goal of guiding me through my grief process unharmed" (2010, p. 160). This voice is represented as a position that engages in frequent intrapersonal dialogue with the grieving self and as constitutive of navigating the hardship of her loss. After several months the pain attacks are represented to have lessened.

> Maybe my body was too exhausted, but more likely I had too much to do. I was busy building a new life. I moved, changed my place of employment, and got used to the new daily life, which was now predominantly occupied with shopping, working and the many things that are part of everyday life rather than licking my wounds, sleeping and writing my diary. (2014, p. 142)

A new context, namely the move from the country to Vienna and entering a new relationship, are new positions and circumstances that prompt a re-evaluation of the bereaved self, thus leading to innovation of the post-loss self. In addition, this self is positioned to have actively taken care of herself by deliberately focusing on key questions supportive of her wellbeing: "What or who is good for me? What makes me stronger? Who or what helped me to survive? And how or what robs me of my energy? How would I like to shape my future?" (2014, p. 146). This evolving self holds the conviction that she will find happiness again (2010, p. 284). As such, the bereaved self is positioned as active processor of her experience of loss by juxtaposing herself in alignment with the voice that deliberately takes care of herself and identifies what and who is supportive of her.

The clown identity: a life-affirming worldview and attitude

An identity that is positioned as pivotal to her recovery and supportive of her wellbeing is the clown self. Pachl-Eberhart positions herself and her husband as clowns, and she portrays their meeting at one of his performances as love at first sight (2010, p. 14). After the accident her husband died immediately, but her seriously injured children were brought to the hospital where Pachl-Eberhart worked as a *Red nose hospital clown* two days a week, and at the exact time when she is taken to the hospital by friends to see her children, she was meant to work there as a clown. At this moment, Pachl-Eberhart's grieving self forms an alliance with her clown self, who expresses her gratitude that a clown colleague greeted her (2010, p. 42). Seven of her clown friends support her when she turns off her son's life support a few days later. At her request, they turn up in costume, first to her son's hospital room, and later, fifty dressed-up clowns help her to turn the funeral into a bright, light occasion, which Pachl-Eberhart positions as a "soul-celebration". The clowns are framed to have put their own grief on hold in order to support her (2010, pp. 79–80). The clown colleagues and friends are external others who cushion the hardships of turning off life-support and are instrumental in making the funeral a celebratory occasion that expresses Pachl-Eberhart's gratitude and love for her family. According to Schmid, meaning making is more important than happiness. He sees not only befriending the self but close and caring friendships with others as an important path to finding meaning (2015).

The grieving self turned to the clown identity to adapt its knowledge and worldview to the altered circumstance of multiple loss. One of the ways in which this identity is framed as being supportive of her grief is the ability of the clown to focus her attention in the present moment: "Being in the moment very much carried me" [through my grief] (Judi1783, 2013). Her clown identity and training are positioned as instrumental in moving through her grief, in particular, the clown's ability to head into the unknown and to trust that everything that she needed would show up (2014, p. 33): "Clown rule number one: No plans. No ideas. Be prepared to be surprised and make the best out of each situation", and she represents herself to have been ready for her path into the unknown (2010, p. 44). The bereaved identity asked herself, "what does a clown do who loses her family?":

> *Of course, he cries. Uninhibited. … he welcomes everything that he finds. Even if it is not what he was looking for. … Naturally the clown fights for what is important to him. … but it is possible that soon the clown will laugh again. He finds the laughter, because he changes his perspective and discovers what is precious.* (2010, pp. 113–114, emphasis in the original).

Pachl-Eberhart is able to find laughter, and she works as a hospital-clown again after five weeks, albeit as a new clown persona (2010, p. 300). The alliance of the grieving and the clown selves are framed to have made it possible to experience pain and joy side by side, and to embark on the path from grief to happiness. On this path, Pachl-Eberhart portrays herself to dig deep and not to shy away from existential questions, asking God: "Tell me, why wasn't I in the clown-bus, when the accident happened? Why did you assign me the fate of being the survivor?" (2010, p. 149). By way of answering herself, the grieving self holds the worldview that there must have been a good reason why she wasn't in the bus (2010, p. 163). Rather than positioning herself as a victim, she actively rebuilds her life:

> I did not want to be a "broken woman". Pursued by fear, I did everything to keep myself together as a whole. My mind ran at full blast searching for a new identity, a new purpose, the meaning that connects everything, past, present and future, happiness and pain. ... I found practical, creative plans for my life. (2014, p. 238)

Like Vujicic, she positions finding meaning and a purpose as pivotal to creating happiness. Schmid concurs with the importance of meaning-making and positions it as central to the art-of-living. He contends that all questions about life, such as "what is important, what is happiness, what is right", are permeated by one other overriding question: "what is meaningful?" Exploring meaning, he states, is a question of connections. Meaning has to be discovered, it is not prescribed, and the act of reflection and interpretation is vital to the process of discovering connections and relationships, both in relation to the self and to others (2008, pp. 210–213).

Pachl-Eberhart positions her loss as constitutive of her post-loss selfhood:

> One day, lying in the grass, I understood that what makes me me today cannot be described with the phrase in spite of. No, I am not in *defiance*. Not *despite* my family's death am I the person today who I am. Not *regardless of*, but *in respect of* my fate. Not *even so*, but *also and especially because*. (2010, p. 146, emphasis in original)

This position shows itself to be embracing rather than resisting and fighting her loss. Again, there is a parallel to Vujicic, who also stresses that his happiness was not created *in spite of* his disability, but because of it. Both authors create their wellbeing in alliance with their adversarial life experiences rather than in defiance of them.

Conclusion

The philosophical perspective of the art-of-living positions self-knowledge as "indispensable" and "narrative self-knowing" as a modern way of undertaking self-reflection and dialogue (Dohmen, 2003, p. 365). It views personal writing as a useful tool for self-reflection (Schmitz, 2016, p. 16). This research of Vujicic's and Pachl-Eberhart's autobiographies investigated the authors' portrayal of their individual art-of-living in the face of adversity. In particular, both convey the attitude that Schmid conceptualises as being at the core of the art-of-living: an attitude that accepts and embraces life in all its polarity, including tragedy, grief and unhappiness.

This research indicates that the dialogical self is capable of renewal in the face of extreme adversity to create enduring happiness and wellbeing through the process of innovation and juxtaposing positions. Pachl-Eberhart describes her interpretation of happiness: "What else could matter at the end of our journey of grief but happiness? I love this word. Happiness" (2014, p. 219). By this she does not refer to the small moments of happiness that she opened herself to from the very beginning, such as suddenly seeing a rainbow. She signifies the "big, authentic, real happiness. Not the type of happiness that comes accidentally and then flies off again, not happiness that is on loan, but happiness ... that is enduring and dependable" (2014, p. 220). This position has contemplated and identified her selfhood as happy:

> I have reached a point where I can say without hesitation that I am truly happy. It took a long time. ... about five years. Does that matter? I don't think so. Seeking happiness is a serious business. To create it with care does take a lot of time. (2014, p. 222)

Both authors positioned and repositioned their disabled and grieving selves respectively to consciously focus on gratitude for what they have or had, rather than what they lacked or lost, and both deliberately put their attention on positive things when they struggled emotionally and met their painful feelings without repressing them. Importantly, the authors position their happiness not in spite of their extreme challenges, but as a result of them. The tone of the texts are life affirming, but not as unambiguous, rather, acknowledging that life is not continuously happy and that difficult moments or experiences are part of life, including a life that is fulfilling. In *Unstoppable*, Vujicic dedicates one chapter to narrating a time in his life as motivational speaker when a short-term cash-flow problem led him into a cycle of despair out of proportion with the actual incident, and he positions this time as a meltdown and as the most challenging in his adult life (2012, pp. 29–50). Further, Vujicic keeps a pair of shoes in his wardrobe as a sign that he is open to experiencing a miracle (2011, p. 57). And Pachl-Eberhart does not frame happiness as simplistic and black-and-white but acknowledges that in her everyday life she has moments of unhappiness and exhaustion, but positions her overriding feeling as happiness (Stöckl, 2018). This research might open up the possibility of working DST into the process of autobiographical writing, for example in clinical practice, in order to lead to the positioning and repositioning of identities which can create happiness and wellbeing, in particular after adversity.

The psychologist Bernhard Schmitz suggests that the philosophical concept of art-of-living has not been investigated much in psychology and argues that it is "a means to achieve happiness" that can be used to predict well-being (2016, pp. 14+17). His research is grounded in the notion that the art-of-living can be learned "by almost everyone", stating that his research team has demonstrated that intervention strategies can be designed to "enhance art-of-living and happiness for different groups" (2016, p. 19). Future research into the art-of-living could include a cross-disciplinary approach, linking philosophy, psychology, narrative psychology, and autobiography scholarship.

Notes

1. All translations are undertaken by the author of this paper
2. https://www.youtube.com/watch?v=Gc4HGQHgeFE

Disclosure statement

No potential conflict of interest was reported by the author.

References

Dohmen, J. (2003). Philosophers on the "Art-of-Living". *Journal of Happiness Studies, 4*(4), 351–371.
Hermans, H. J. (2001). The construction of a personal position repertoire: Method and practice. *Culture & Psychology, 7*(3), 323–366.
Hermans, H. J. (2004). The innovation of self-narratives: A dialogical approach. In L. E. Angus, & J. McLeod (Eds.), *The handbook of narrative and psychotherapy: Practice, theory and research* (pp. 175–191). Thousand Oakes, CA: Sage.
Hermans, H. J. (2008). How to perform research on the basis of dialogical self theory? Introduction to the special issue. *Journal of Constructivist Psychology, 21*(3), 185–199.
Hermans, H. J., & Gieser, T. (2012). History, main tenets and core concepts of dialogical self theory. In H. J. Hermans, & T. Gieser (Eds.), *Handbook of dialogical self theory* (pp. 1–22). New York, NY: Cambridge University Press.
Judi1783. (2013, September 13). *Trauer[bewältigung]: Barbara Pachl- Eberhart, Zusammenschnitt*. [Video file]. Retrieved from https://youtube/um0ccyP7QPg.
LeMahieu, D. L., & Cowley, C. (2018). Philosophy and life writing. *Life Writing, 15*(3), 301–303.

Pachl-Eberhart, B. (2010). *Vier minus drei: Wie ich nach dem Verlust meiner Familie zu einem neuen Leben fand*. München: Integral.

Pachl-Eberhart, B. (2014). *Warum gerade du?: Persönliche Antworten auf die großen Fragen der Trauer*. München: Integral.

Salgado, J., & Hermans, H. J. (2005). The return of subjectivity: From a multiplicity of selves to the dialogical self. *E-Journal of Applied Psychology: Clinical Section*, *1*(1), 3–13.

Schmid, W. (2007). *Glück*. Frankfurt: Insel Verlag.

Schmid, W. (2008). Mit sich selbst befreundet sein. *Aufklärung und Kritik*. Special Issue, *14*, 209–219.

Schmid, W. (2015). Wilhelm Schmid: Kein Glück ohne Unglück. [Video file]. Bildungs TV. Retrieved from https://www.youtube.com/watch?v = DNGafut8TL4.

Schmitz, B. (2016). *Art-of-living. A concept to enhance happiness*. doi:10.1007/978-3-319-45324-8.

Stöckl, C. (2018). "Ö3-Frühstück bei mir – Spezial" zum Muttertag. [Radio Broadcast]. 13 May 2018. Vienna: Ö3. Retrieved from https://files.orf.at/vietnam2/files/oe3/201819/oe3_fruehstueck_180513_597448.mp3.

Vujicic, N. (2011). *Life without limits: How to live a ridiculously good life*. Crows Nest: Allen & Unwin.

Vujicic, N. (2012). *Unstoppable: The incredible power of faith in action*. Crows Nest: Allen & Unwin.

Vujicic, N. (2018a). *Attitude is Altitude*. Retrieved from https://www.attitudeisaltitude.com/.

Vujicic, N. (2018b). *Life Without Limbs*. Retrieved from https://www.lifewithoutlimbs.org.

Vujicic, N., & Vujicic, K. (2014). *Love without limits: A remarkable story of true love conquering all*. Crows Nest: Allen and Unwin.

Renewal and multi-voicedness: in search of narrative conscience

Michiel de Ronde

ABSTRACT

In this paper, the theme of multi-voicedness in the search on the path of life is explored. It will be argued that we, as human beings, not only live within an ambiguous world, but also have to deal with an ambiguous self. In the quest for happiness it is the challenge to acknowledge all inner positions, each with their own perspective on reality. The article describes a narrative method based on the archetypal biblical story of the prodigal son. By retelling this three times, as three symbolic stories, namely that of the youngest son, the oldest son and the father, the participants are invited to probe their lives from different sides and accept all inner voices that speak.

Phoenix

Flame in me, flame again;
heart in me, be patient,
double your trust –
bird in me, let me spread my wings anew
now still tired and blue
oh, wing up now out from the burned branches
and let thou courage and thou speed not fail
the nest is good, but the sky is ever more spacious.
H. Marsman (translation R. Lengelle)

Renewal

The image of the Phoenix comes from Greek mythology: the eagle is cleansed by the fire, rises out of its own ashes and through this process of fear and pain gets a second life. The Dutch poet Hendrik Marsman (1946) takes this image and uses it in a poetic dialogue with the self. He (more accurately the lyrical "I" in the poem) speaks to the flame in himself. He calls to his heart to trust more. He seeks the bird in himself, that part that can spread its wings and fly … In the poem the final question is about what the good life means: " … the nest is good, but the sky is ever more spacious." It seems that the good and happy well-being can be a barrier of what is better. In the dialogue that the poet has with himself, the choice must be made for the sky with its uncertain spaciousness, which means the good nest must be left behind. That is why it is inevitable: the flame must arise. The poet knows this and speaks, like the Phoenix, to himself to tell himself that he must go through the fire in order to re-find this spacious sky.

The Phoenix is the mythological representation of renewal and transformation of the person. How can a person arrive at a new perspective on the world and a new understanding of himself within that world? And what is its relation with the sens of well-bing? I present here a form of guidance that helps people with the use of an archetypal story to enter into a dialogue with themselves which will offer deepening and expansion surrounding life choices.

Moods

The constrictive pressure that accompanies having to make life choices is something I know well from personal experience. There have been several periods in my life that the intense fear of not knowing gripped me. I would wake up in the morning with an unhappy feeling of pressure and threat. In those moments, at dawn, I was afraid of everything. There was not really a fear of something in particular but often the fear would attach itself to something at random. I was as scared as a child for the things I had to do on those days and was worried about the judgments of others about what I had done the day before or what I should have done. My fearful feelings could be about how my children were doing, or the state of my house and the payment of my mortgage. Things appeared in dark and threatening guises and I felt pressed down by the weight of things that seemed hard to bear. I would have gladly hidden under a blanket but I knew that was not wise and would not bring good things. In those moments it was a matter of focusing on duty, following a routine and enacting some will power: the discipline not to let myself be lead by this gnawing feeling, but rather to be lead by the steady rhythm of the days and the duties that were waiting for me.

What was, however, interesting was that the dark images could suddenly shift. Often enough I experienced that once I was up, had eaten breakfast and closed the door behind me, climbed on my bike to ride to work, that suddenly another feeling took hold of me. I breathed the fresh air in. I heard birds singing among the green and the wind rustling the leaves of trees along the way. The image shifted. The dark threat of fear simply made room for a joyful feeling of happy wellbeing and contentment. The soft light of the morning sun, the welcome coolness of the dawn and the friendly cooing of doves told me that life is good and that the world was abundantly beautiful.

Time and again I was surprised by this shift in perspective, which happened to me: two moods, two incompatible ways of being and attitudes with which I met the world and, as an inevitable result, with which the world presented itself to me. The transition from the safe nest to the spacious sky that the Phoenix experiences turned out not to be a singular experience but was a repeating process that I had to go through again and again each time. The constriction surrounding complicated life choices happened in a repetition of perspective shifts between opposites: light and dark, life and death. Despite the repeating nature of the experience, I ask myself now, is there a corresponding development and growth? Was there an increase in insight and acceptance about the path that I went and go?

The psychology of perception

The well-known image of the older and young woman, that can be found in almost every introductory chapter of a psychology text about human perception, has because of the experience I describe gained for me new existential depth (see Figure 1). From a quaint example of the illusory nature of our perception, it became a fundamental illustration of the multiplicity of the world in which we live and from the always present ways to shift perspective. Systematic reflection on this and similar images provides insight into certain recurring laws that exist in the psychology of perception.

First, we see the overall picture. We see it as an image of an old woman and a young woman. From the whole we can point to a part: this is the chin of the old witch, or this is the nose of the pretty girl,

Figure 1. Image of the older and young woman.

but always there are parts of the whole. The other who we show the image to, by pointing to these parts of the not yet discovered second image, suddenly sees the whole (and therefore also the individual parts that we're pointing to). Also in the personal experience I described upon waking, the world shows itself to me in the same way, as a whole, as threatening and pressing, or as the shift to happy and inviting. Our perception provides us an image of the world that has the quality of a "Gestalt", a coherent whole that is more than the sum of its parts (Köhler, 1951). The transition from the one image to the other is captured in the German-English term, "Gestaltshift".

Second: the separate parts get their meaning in the relationship to one another and their place within the whole. Because in the one image the white is the naked throat of the young woman, the black line inevitably looks like a necklace. At the moment the white shifts and becomes the chin of the old woman, that same black stripe immediately becomes her mouth. The one is inseparable from the other. The sense and meaning of the things around us are part of a total image and through the whole their separate value and meaning. If I get up with a feeling of anxiety that will express itself in how I look at myself in the bathroom mirror and also in the way that breakfast will taste. The parts get their meaning from the whole.

The third law flows from the first two: we can only see one of the images of the world at a time. At a certain moment we see either one or the other. For those who do their best to see both at the same time, they will notice that the images start shifting quickly from one to the other and back again. We can actively seek the old woman if we primarily see the pretty girl or vice versa. Maybe – if it works at all, and it will likely cost a lot more effort! – one can also make an effort to shift from a somber mood to an optimistic one.

Personal experience

The reflection above illustrates the multi-dimensional nature of the reality in which we live. The world shows up in a variety of forms. The example of my anxiety at the start of the day takes us to another law of the psychology of perception: the perceiver and what is perceived are one and the same. They influence each other reciprocally. On the one hand my sombre mood coloured the world in tints of grey; on the other hand the things in the world around me were what contributed to the shift in how I experienced myself. It is not only the outside world that shifts, but also my inner world, which takes shape in my self-experience, that changes in quality. It turns out we ourselves, just like the image, are an ambiguous reality. The multidimensionality of the world and the multidimensionality of the self are directly related to one another. Goethe (1982) said it already: "one person, many dimensions". And just like the shifting of the image, we can actively seek that inner shift. The poem by Hendrik

Marsman, that is the motto of this chapter, demonstrates a kind of self-talk where such an active effort is tested.

This insight that we as people do not only encounter a multi-dimensional world but a multi-dimensional self has been argued and described in psychology by various authors. Freud (1932) of course describes the topological model of the psyche, where the conscious I can be overtaken by the instinctive drives of the "Es" and the norms of the Über Ich. Jung (1995), inspired by Freud, considered psychological complexes (for instance the so-called archetypes) as kind of independent entities in the (un)conscious. Psychological growth means learning to accept and integrate these more or less autonomous parts of the self. Jung's thinking on this later lead to the development of the popular *Voice Dialogue* approach, where human personality is viewed as a collective of voices or sub-personalities (Brugman, Budde, & Collewijn, 2010; Stone & Stone, 1989). Related to this way of thinking, but coming from a different source is the school surrounding the theory of the Dialogical Self (Hermans & Hermans-Konopka, 2010; Vassilieva, 2016). In this theory, the human personality is conceptualised as a dynamic multiplicity of I positions, that each bring their own perspective to the world

Exploring the multi-voiced

In my coaching practice, I am in search of working methods that can contribute to the exploration of these multiple perspectives in which we live and that provide support for dealing with the life questions that can occupy our minds. I asked myself, can I offer useful methods that help my clients and my students to engage in a dialogue with themselves like the poet does in "Phoenix", a dialogue on the level of the self-experience? Can I help them explore and see in an honest way the question about the good and happy nest and the spacious sky by bringing this question close to the level of experience?

A narrative form of coaching: the story of the father with two sons

As a child of orthodox-protestant parents, I was raised with Bible stories. I not only heard them Sunday at church and during Sunday school, but these stories were also read at home around the table at mealtime. In many occasions these stories were about good and evil, about what was worth pursuing or what was objectionable. There was the paradise and the fall and there were Cain and Abel, Abraham and Lot, and Jacob, Esau, David and Saul. Each time one of the two aspects or characters was either gifted and rejected, like positions of light and dark. In a similar way there were the stories from the New Testament of Jesus, stories that we refer to as the "likenesses". One of these parallel stories from the New Testament is the story of the two sons, which is also called the story of the Prodigal Son. It is a story full of symbolic drama. One day the youngest son says to the father, "father can you give me the inheritance that would be mine when you die? I would like to go into the world, because the nest is good but the sky ever more spacious ... "

It is this archetypal story I use in various forms to coach people as they reflect on their own life journey, as they wonder about the question of what they must do with the inheritances they have received (de Ronde, 2011). It is also this story that has helped me with big life decisions – helped me to see and respond to the question of the good in an honest way, and strangely enough, helped me to transcend the polarising contradictions of good and evil. The good turns out to not always be good and happy and the so-called evil is sometimes the best choice for well-being.

In the core of the approach that I've developed here, comparable to the image of the old witch and the young woman, one is helped to see multiple stories in the single story and to take each of those stories and apply them to one's self. I ask those I work with – often I work with a group, but sometimes also an individual – which theme is pertinent for them at the moment (step 1). Then I invite them to read aloud the story I've described above (step 2), an ancient history that after the reading I will repeat three times because in this one story there are three stories hidden (step 3). After each re-telling I invite the participants (sometimes with some prompting questions) to apply the gist of the story to their own life theme (step 4). Finally I ask them to if they can learn something from this in the context of what they will be doing tomorrow (step 5).

Reading aloud

This is the text I use as I am reading:

Lukas 15: 11–32

11 And he said, A certain man had two sons: 12 And the younger of them said to his father, Father, give me the portion of goods that falleth to me. And he divided unto them his living. 13 And not many days after the younger son gathered all together, and took his journey into a far country, and there wasted his substance with riotous living. 14 And when he had spent all, there arose a mighty famine in that land; and he began to be in want. 15 And he went and joined himself to a citizen of that country; and he sent him into his fields to feed swine. 16 And he would fain have filled his belly with the husks that the swine did eat: and no man gave unto him. 17 And when he came to himself, he said, How many hired servants of my father's have bread enough and to spare, and I perish with hunger! 18 I will arise and go to my father, and will say unto him, Father, I have sinned against heaven, and before thee, 19 And am no more worthy to be called thy son: make me as one of thy hired servants. 20 And he arose, and came to his father. But when he was yet a great way off, his father saw him, and had compassion, and ran, and fell on his neck, and kissed him. 21 And the son said unto him, Father, I have sinned against heaven, and in thy sight, and am no more worthy to be called thy son. 22 But the father said to his servants, Bring forth the best robe, and put it on him; and put a ring on his hand, and shoes on his feet: 23 And bring hither the fatted calf, and kill it; and let us eat, and be merry: 24 For this my son was dead, and is alive again; he was lost, and is found. And they began to be merry. 25 Now his elder son was in the field: and as he came and drew nigh to the house, he heard musick and dancing. 26 And he called one of the servants, and asked what these things meant. 27 And he said unto him, Thy brother is come; and thy father hath killed the fatted calf, because he hath received him safe and sound. 28 And he was angry, and would not go in: therefore came his father out, and intreated him. 29 And he answering said to his father, Lo, these many years do I serve thee, neither transgressed I at any time thy commandment: and yet thou never gavest me a kid, that I might make merry with my friends: 30 But as soon as this thy son was come, which hath devoured thy living with harlots, thou hast killed for him the fatted calf. 31 And he said unto him, Son, thou art ever with me, and all that I have is thine. 32 It was meet that we should make merry, and be glad: for this thy brother was dead, and is alive again; and was lost, and is found.

I consciously choose a version from the King James bible. This thick official and leather-bound book and the formal language in this volume, together create the transitional space (Winnicott, 1971) in which the holy and the magical come together. Because the reading/telling ends up in a mythical atmosphere, the listener is also invited to view his/her own life in the light of the mythical, where we are part of something bigger than ourselves and where we are still charged with finding our own way. By doing this, we become able to also discover and affirm the myth of our own lives as a form of a meaning-giving perspective (Bohlmeijer, 2007).

First re-telling: the story of the youngest son

After the presentation of the story there is a momentary silence. Those present are still captured by the words and remain in the atmosphere of the story. The ending affects them. Then I step out of my narrator's position and say to them,

Actually the story here contains three stories, one about the youngest son, one about the oldest son and one about the father. And each of these three stories are stories about ourselves; they each form a part of our own story. I will tell you each story again and I will do it in a way that I put myself in the shoes of each of the characters, the youngest son, the oldest son and the father. As if I am each of them.

As I take a step in order to move into a new position in the space, I begin …

"I, youngest son, as I look back, of course I knew that I hurt my father. Who does that – demand to get their inheritance before their father is even dead? And then even spends the money on far-away journeys and does so extravagantly. I did it. Now I'm ashamed but then … It was a brash deed, and at the same time … I couldn't stop myself. I had to do something. My father's home had become too stifling for me. I couldn't breather anymore. I felt stuck. I had to get out. To seek life, to find my way and see the world. It was a compulsion too strong to resist. I believe I would do it again in the same situation … Of course I knew when I'd gone and had one party after another and treated my so-called friends, that it was not life. Still, I did it anyway. Until I had used up my inheritance and there was no money left. Then I was confronted with myself and I had to take care of myself. I did the lowest of low to even earn a little bit and to get a bit of food. The pigs that I took care of had it better than I did. They could at least fill their bellies; I went hungry.

This was the turning point. I began to ask myself the questions that I had not asked before. Up until then I had merely followed my drive, more or less blindly and without thought. Now I engaged in a conversation with myself. I spoke to my heart and said to myself that it was I who had lost my rights and thrown away my life. I understood that my own deeds had led me to this moment. I realised that I myself was responsible for the path I had gone and the sorrow I had caused my father. The latter pains me the most. My thoughts went back to the past. Something I had never done before, I did now. I thought about what the life of my father's servant is like. Those who live without freedom on the parental property. That is what I shall do, I spoke to myself: I shall carry the debt and the bondage of the servant, now that I have misused the freedom as a son. That is how I returned, back to my parental home, back to the property of my father.

Now, so many years later, it is still difficult for me to say what happened afterwards. The most difficult thing in fact turned out to be accepting my father's love, without me having anything to offer in return. I had reasoned in terms of guilt and punishment. I would take on the slave position as a consequence of my choices. What happened instead, I could not have predicted I was dressed anew as son with coat and ring. A feast was held. To receive all this was actually the most difficult. The only reason for me not to refuse was that I didn't want to hurt my father yet again, I did not want to push him away. In all honestly, this is still a daily struggle for me: to receive without giving in return. That is how I live now. I comfort myself with the thought that I have been able to take a bit of pain away from my father. Maybe that is what I do in return: carry the pain I caused him".

After this first re-telling, there is a silence again. Then I ask a number of questions which have a common focus. What do you recognise regarding your theme based on the re-telling as it pertains to:

• hurting another in order to be yourself;
• admitting to guilt and to acknowledge your own shortcomings;
• accepting your own shadow side;
• receiving from others, without being able to give back?

Second retelling: the oldest son

Once again I move to a different spot in the room in order to emphasise physically that ever person takes in a different position in the story and with that has their own perspective in the world.

"It is not easy for me as oldest son to, after so many years, revive this story. It rips open old wounds again. Maybe this is good because the wounds have never really healed in me. When my younger brother wanted to leave those years ago, I was not really surprised. He was different than I was. He was impatient and didn't tolerate the quiet well. I found there was satisfaction in tending to the land and the animals. That was not the case for him, he always sought adventure, always the path of uncertainty. I can't really blame him. I know him. He is the way he is. But at that dramatic moment when my brother demanded his inheritance and my father gave it to him with tears in his eyes, everything changed. I did not only lose my youngest brother, I also lost my father. From that moment on he stood waiting. As of that moment his heart was no longer with us, but always with my brother, far away. He carried his smart in silence, but despite that, I was also an inevitable part of that sadness.

And I? I realised that the care of the property rested on my shoulders. Now that my brother was gone and my father could do nothing else but think of him, I had to make sure that everything kept going. What I had done before with pleasure now became a duty. The work that I had once done, before my brother left, with joy and natural commitment, I now did with the pressured feeling of responsibility. Did I have to make up for my brother errors? Did I, in the knitted weave of life, have to pick up the thread that my brother had dropped? This is not the whole story. Before this happened, I had already had to take care so that my father's life and the life of the property could continue. Even if it had been brought to a halt; even it had actually become lifeless and dead. It is difficult to put into words, but that heavy feeling of pressure is something I can still easily conjure up. I only have to think about it and it's there again. That lonely feeling, of carrying the world on your shoulders, without being seen.

And then there was the day that my brother suddenly returned home again. All of a sudden there was life, all of a sudden there was joy. Bang and there was a party. Something broke inside me then. What broke was the part that had been under pressure all those years. I couldn't do it! I couldn't!

My father came to me. He pushed me to join in, to be happy. But he was a stranger to me. All those years I had hoped for a look from him. Not only the sad eyes that looked out towards the other child, but eyes that saw me. That was in the end what I had always done it all for. But never had I received. And at that moment, when my youngest brother was there again, it was expected of me that I would be joyful and take part in the feast. I could not do it. It was the first time that I disobeyed my father. I shouted at him, I swore. I congratulated him on my unfulfilled longing. I turned around and left.

When I look back at that now? I would do the same, in the same way. Not because I am proud of it. On the contrary, it is associated, even for me, with feelings of shame and blame. But still, I would do it again, because it was an expression of my heart, because it was an honest response to my pain and yearning. Maybe my path resembles that of my brother. The step to freedom may indeed not be possible without guilt. I hope that my father understands it a bit".

Also now, after this telling, there is a moment of silence and I take another little step sideways, use prompting questions on the topic: What do you recognise in the oldest side in relation to your theme:

- being obedient because it is proper to do so;
- carrying responsibility when others neglect to do so;
- experiencing the pain of not being seen;
- being honest with/for yourself and towards others when it's important to do so?

Third re-telling: the story of the father

Once again I move to a different spot in the room. The story of the father that will be told now, after all, requires a different perspective once more.

"The path through life is mysterious. If I have learned anything as a father in the many years of my existence, it is that you are most vulnerable with your most precious possessions. The word possession turns out not to apply as what is your most precious possession can be lost at a moment's notice.

That my youngest son wanted his inheritance, that indeed was about property and possessions. Of course I could have refused. Of course I could have become angry and told him the truth. However, I knew if I had done that that I would not have be able to keep him near me. On the contrary. Maybe he would have stayed on the family property for a while but he would have hardened and closed himself and I would have lost him. That I knew only too well. He saw the sadness in my eyes, when I gave him his part of the inheritance. I saw that he saw it. I also saw that he didn't allow himself to realise it. Then he turned around and left. I said to myself, this is the path of the father: learning to let go; letting the child go. I can tell you that it was not easy. He was my youngest. In all his busyness and restlessness, he was my favourite. I had to conquer myself in order not to hold on to him. To let him go into the world, whatever that would bring. I am still proud that I was able to do that and that I did it. But of course my heart stayed with him. I waited. Every day I waited for the moment that he would appear on the horizon again.

I realise now, looking back, that I failed my other son, that I have denied him so much. I took it all for granted. He cared for the farm, for the property. Everything that was mine was his too. But only now do I realise that I did not look him in the eyes. Now I realise that my heart was not with him. I was a father, of course I was also the father of my oldest son, but still, my eyes were always pointed to the far horizon, there at the ends of the earth, to where my youngest had disappeared …

And when he returned, destitute and damaged, poor and naked, my heart jumped for joy! That is when my happiness returned. I could not help but to put on a great feast. He was dead and found alive. I myself had been dead and received new life. I embraced him, I kissed him, I dressed him … my son, my dearest.

And, how sharp are the thorns of life, when he returned, that very night, I lost my other child. He did not want this. He could not bear it, he told me. I saw it in his eyes. He looked at me. He saw my joy. His face hardened … in one moment, I saw how it actually shifted. He looked at me with the eyes of a stranger. He turned around and left.

Yes, and now? I love my children. Both equally. They have both hurt me. I realise now more than ever that I also let them down. Especially my eldest. It is the law of life. Children must free themselves from their parents. The feast for my youngest was the moment for my oldest to go. And now? Do I also wait for him, until he turns to me again … or is something else demanded of me now? Shall I seek him perhaps, in order to tell him what I realise now? Yes, that is what I shall do. I shall stand up and go to my oldest … The story is not yet done".

The story of the father also requires a moment of silence and is followed by a number of questions that can likely be related to one's own life themes. What might you learn from the perspective of the father about:

(1) letting go of what is your own and with which you are deeply bonded;
(2) carrying the pain of life through facing a sense of loss;
(3) the realisation of having paid too little attention to another's life or pain;
(4) taking initiative in opening things up despite that this is difficult.

What the process yields

The conversations (mostly done in pairs) that happen in response to the three stories are touching and intense. Everyone it turns out has something of the youngest son, the oldest son, and the father in him/herself. Just like the characters in the stories, the participants are able to recognise the tensions in the fundamental drives of connection, responsibility, and care for others on the one hand and freedom, autonomy and personal development on the other. These drives can be conceptualised as voices in the landscape of the self, that are in conversation with each other.

The tellings here break through the idea that one position is the right one and the other position is less right. The path of the three characters in the story, the youngest son, the oldest son and the father all have light and shadow sides. Primarily the telling contributes to the acknowledgement that the good choice, in many cases, is inevitably paired with a sense of guilt and having fallen short. In his beautiful book "Fear and Trembling" (Kierkegaard, 1983) the existential philosopher Kierkegaard, for instance, shows that the existential lies beyond the moral. No law book, no ethical code, no father, no mother or life partner can determine for us what the right life choice is. That is only something each person can determine for his/herself. It is precisely this realisation that is recognised and confirmed by this telling. The way in which the choices described are made, is not only beyond the moral, but also beyond the rational. The perspectives of the three characters are just like the images in that one picture, they each form a whole, a "Gestalt". Every perspective carries a world along with it. Sometimes one world is in the foreground and at another time, another world/perspective is dominant. By acknowledging every perspective from the inside out and by accepting it as a realistic possibility, the multiple I-positions of the participants can be seen and affirmed in their individual existence. It is precisely this acceptance (of the reality of the different orientations) that offers the foundation for taking action. In the dialogue between the different positions Verhofstadt-Denève (2007) sees the path to individuation; from that dialectic newly integrated perspectives can come to be.

Renewal and multi-voicedness

Every person deals in their own way with the inheritance they have been given in life. The archetypal story of the father and the two sons can be seen as a symbolic narrative about an universal human question. The path of the youngest son who demands his piece of his inheritance in order to live in luxury is a different one than that of the oldest son who suffers in silence taking responsibility for caring for home and hearth. The conversations that emerge among the participants in response to the stories are about different perspectives that we carry within us and about the perspectives with which we approach the world and respond to the life questions that appear. Each story, each perspective creates an individual world. We might ask ourselves where the renewal is to be found when we take an old story from an old religious tradition and use it in this symbolic way. My answer to that question is varied.

The first thing the process of telling the three stories yields is that we become aware of the multiplicity of our perspective. We can identify with all three versions because in a certain way all three of these perspectives exist within us. The result of this is the deconstruction of the dominant discourse of dichotomous thinking in "or-or" terms. The process is not about answering the question which of the three characters we are most like, rather the challenge is to recognise in ourselves the presence of each of the three I-positions. The first form of renewal is therefore the recognition of our multi-voicedness and with that overcoming our tendency to deny or repress those parts of ourselves we tend to condemn (According to Hermans & Hermans-Konopka, 2010). The transformation is not so much about finding a new and happy perspective. The transformation is made of the fact that we can change perspective, that we can look at things in a multiple way.

The second thing that this process yields is that the mere recognition of the differing I-positions in the landscape of the self leads naturally to development within those I-positions and in their relationships to one another. That which is stuck, is made dynamic. In the story a movement takes place within the character of the youngest son: his yearning for freedom is complemented by his renewed longing for connection. The development in the character of the oldest son moves in the opposite direction: his initial dedication and care develops towards more autonomy and letting go. Even in the case of the father – in most interpretations the prime example of goodness and simplicity – it seems the loving heart has its shadow side. Every person who is willing to look at the different aspects of him/herself will discover that good and bad, light and dark, desirable and objectionable are much less unambiguous than assumed. Perhaps individuation, becoming, is actually the acceptance of the side of us that we tend to condemn (Jung, 1995a). Light cannot exist without darkness, and the dark cannot exist without light. In the image of the old woman and the young woman, the dark and white together make the two "gestalt". Well-being requires acknowledgement of both.

The third thing I see that this process yields – extensions of the first and second yields – is an expanding "courage to be". Tillich (2004) describes this as a form of self-acceptance, whereby we simultaneously accept that our own life choices are filled with shortcomings and guilt and we also accept and recognise that this is personally the right path. Listening to the story in this triple form offers the freedom to confirm one's own life path and to make life choices, including the full recognition that there is always a shadow side connected with this.

Light and shadow

The core question about what a good life is appears in the shape of a dialogue with the self. That question only truly becomes a question for us when we dare to ask it of ourselves. What should I do? What is for good for me, the nest or the universe? What do I allow myself to be driven by, freedom and variety or safety and predictability? Which values determine my sense of direction: attention for the other or a focus on myself? What drives me in my choices: conforming and loyalty or confrontation and autonomy? Those who are faced with the inevitable challenge of making life choices and must answer unavoidable ethical-existential questions and with that experience the pressing fear of not knowing, are in my view helped by having an approach that help one recognise and feel from within the differing perspectives and their value and reality and to feel the multiple I-positions in the landscape of the self. That is in my view the core of the narrative conscience: by experiencing the movement of an archetypal story about an universal human theme in which every position has a right to exist and is genuinely accepted as a reality in its own right, we are able to shift the dichotomy of good and evil towards a dialectic between light and shadow, a dialogue between love and fear, between freedom and guilt, happiness and sadness. The triple reading of the story of the father with the two sons helps me to transcend the black-and-white thinking – to see it as too limiting a perspective. Just like in the black and white image of the old woman in which the young woman is enclosed also, there are many configurations to be discovered in our the light and shadow of our minds, each with its own meaning and perspective on the world. In good, evil is also often enclosed; in what is seemingly objectionable the gain of the good is often present (Jung, 1995b). Who will burn the good nest? Indeed, the poet of the Phoenix calls, "Flame in me, flame again ... "

Acknowledgement

My thanks goes to Dr. Reineke Lengelle for her assistance with translation of this article.

Disclosure statement

No potential conflict of interest was reported by the author.

References

Bohlmeijer, E. (2007). *De verhalen die wij leven. Narratieve psychologie als methode*. Boom: Amsterdam.
Brugman, K., Budde, J., & Collewijn, B. (2010). *Ik (k)en mijn ikken*. Zaltbommel: Uitgeverij Thema.
Freud, S. (1932). *Neue Folge der Vorlesungen zur Einführung in die Psychoanalyse*. Frankfurt am Main: Fischer Verlag.
Goethe, J. W. (1982). *Faust*. Wereldbibliotheek: Amsterdam.
Hermans, H. J. M., & Hermans-Konopka, A. (2010). *Dialogical self theory: Positioning and counter-positioning in a globalizing world*. Cambridge: Cambridge University Press.
Jung, C. G. (1995a). *Libido in transformatie*. Rotterdam: Lemniscaat.
Jung, C. G. (1995b). *Mensbeeld en godsbeeld*. Rotterdam: Lemniscaat.
Kierkegaard, S. (1983). *Vrees en beven*. Baarn: Ten Have.
Köhler, W. (1951). *Gestaltpsychologie: Een inleiding tot nieuwe opvattingen der huidige zielkunde*. Utrecht: Bijleveld.
Marsman, H. (1946). *Verzamelde gedichten*. Querido: Amsterdam.
de Ronde, M. A. (2011). Op zoek naar de alteriteit in onszelf. Bijbelverhalen als bron van wijsheid in begeleiding. *Counselling Magazine, 2*(3), 41–45.
Stone, H., & Stone, S. (1989). *Embracing ourselves: The voice dialogue manual*. Novato, CA: New World Library.
Tillich, P. (2004). *De moed om te zijn. Over de menselijke persoonlijkheid en de zin van het bestaan*. Utrecht: Bijleveld.

Vassilieva, J. (2016). *Narrative psychology: Identity, transformation and Ethics*. London: Palgrave Macmillan.
Verhofstadt-Denève, L. (2007). *Zelfreflectie en persoonsontwikkeling: Een handboek voor ontwikkelingsgerichte psychother-apie*. Leuven: Acco.
Winnicott, D. W. (1971). *Playing and reality*. New York: Routledge.

Minimalist life orientations as a dialogical tool for happiness

Jennifer E. Hausen

ABSTRACT

In contemporary culture, it is natural to think that purchasing and owning the "right" possessions results in happiness. This belief supports our loyalty to consumer society. However, several lines of research demonstrate that high consumption lifestyles and materialistic values are not a trustworthy path to well-being. Instead, materially simpler lifestyles such as minimalism, with a focus on intrinsic values, have been suggested as contributing to happiness and fulfilment. Thus, the present paper exemplifies how individuals adopt minimalism. I propose Dialogical Self Theory (DST) to explain the decision-making processes in the transition from a materialist to a minimalist lifestyle since DST provides a useful framework to explore multiplicity in an individual's self-concept by recognising the self as moving between multiple and relatively autonomous I-positions. Thus, DST can be used to understand how internal inconsistencies, for example between a materialistic and a minimalistic I, are resolved. My elaborations suggest that the dialogical relations of different I-positions serve as a form of self-guidance leading the self to transform into a minimalist. Implications and suggestions for future research are discussed.

Introduction

Happiness and satisfaction with one's life are of great importance to people all over the world (Diener, 2000). In contemporary culture it is natural to think that purchasing and owning the "right" possessions results in happiness. In line with the latter, Kasser (2002) points out that "articles, images, and advertisements on television, radio, highways, and the Internet proclaim how much happier life would be with this product or that image" (p. 52). As we see happy, admired, and successful people with money, the "right" possessions and high status in our day-to-day lives on television and social media (Dittmar, 2008), the belief that people would be happier if they were richer or rather if they possess the "right" objects persists in contemporary culture (Kahneman, Krueger, Schkade, Schwarz, & Stone, 2006). In line with this, materialistic values (money, image, fame) seem to be increasing in Millennials (born after 1982) whereas goals related to intrinsic values (self-acceptance, affiliation, community) are considered less important (Twenge, Freeman, & Campbell, 2012). However, beyond the point of providing food, shelter, and safety, increases in wealth contribute little to an individual's well-being (Kahneman & Deaton, 2010). Moreover, several lines of research continue to demonstrate that materialistic values are associated with diminished personal well-being (e.g. Twenge et al., 2010). If money, wealth, and status do not lead to lasting happiness, what does? If high consumption lifestyles are not even a trustworthy path to well-being, which lifestyle is? Voluntary Simplicity (VS) or rather Minimalism has been suggested as a viable and desirable alternative to higher consumption lifestyles leading to lasting happiness, fulfilment, and satisfaction.

Since minimalism is regarded a niche lifestyle it seems that mainstream society is not set up for simplifying or for society to become more sustainable (McGouran & Prothero, 2016). At the same time, the increasing online media attention given to the practice of minimalism is also thought to reflect the number of people now adopting this lifestyle (Ballantine & Creery, 2010) which highlights that further research is required to better understand this lifestyle concept. Thus, the purpose of the present paper is to exemplify how individuals consider adopting minimalism as a lifestyle. I propose the use of Dialogical Self Theory (DST) to explain the decision-making processes in the transition from a materialist to a minimalist identity since DST can be used to attend to multiple I-positions and the dialogical relations between positions to understand how internal inconsistencies, for example between a materialistic and a minimalistic I, are resolved.

Therefore, the present paper is organised as follows. I will begin with a literature review on subjective well-being and materialism to justify the proposition of minimalism as an alternative lifestyle. Then, after having outlined Dialogical Self Theory, I will exemplary present the transition from materialist to minimalist from the perspective of DST. The paper concludes with a discussion and suggestions for future research.

Conceptual framework

Subjective well-Being

Research on subjective well-being (SWB), commonly referred to as happiness or satisfaction, has become a dynamic field attracting various researchers from a diversity of disciplines like psychology, economics, political science, sociology, and anthropology (Diener et al., 2017) which argues only for the relevance of insights coming along with the study of this construct. Diener (1984) defines SWB as people's overall appraisals and evaluations of their own lives and their emotional experiences and thus relates to how people feel and think about their lives. It is a broad umbrella term that can be divided into affective and cognitive well-being (Busseri & Sadava, 2011; Diener, 1984). Affective well-being (AWB) refers to the presence of positive and pleasant emotional responses (positive affect), such as enjoyment or gratitude, and the absence of negative and unpleasant emotional responses (negative affect), such as anger or worry, to ongoing life events and circumstances (Diener, 1984; Diener, Oishi, & Tay, 2018). Cognitive well-being (CWB) refers to reflective cognitive judgements of life overall (e.g. global life satisfaction), but also of specific life domains, such as job or health satisfaction (e.g. Diener, 1984; Diener et al., 2018). Thus, an individual with high SWB experiences more positive than negative affect and is relatively satisfied with life (Diener, 2000). Given its subjective nature, the elements contributing to life satisfaction and similarly, what events and circumstances are evaluated as desirable, is determined only by the individual (Diener & Oishi, 2000). Further, research suggests that SWB is malleable (Tay & Kuykendall, 2013) as many controllable factors beyond genes like psychological need fulfilment, changing personal circumstances, or societal factors can increase or lower SWB (for a review see Diener et al., 2018).

A growing body of evidence also focuses on the generally beneficial outcomes of high SWB across a variety of life activities. Individuals with high SWB appear to be of better health and benefit from greater longevity (Diener & Chan, 2011; Lyubomirsky, King, & Diener, 2005). Moreover, high SWB improves social relationships as positive affect is positively related to social interaction quality (Berry & Hansen, 1996) and makes people feel more sociable (Whelan & Zelenski, 2012). High SWB also leads to more prosocial behaviour as people are more cooperative, show increased trust (Lount, 2010), and are more likely to volunteer (Oishi, Diener, & Lucas, 2007).

Money and materialism

For some individuals, money becomes the purpose of their lives as they believe that money, or rather what they can purchase with money, will give them what they yearn. This belief is widespread in contemporary culture and is more widely known as materialism. Dittmar, Bond, Hurst, and Kasser (2014) define materialism as "individual differences in people's long-term endorsement of values, goals, and associated beliefs that centre on the importance of acquiring money and possessions that convey status" (p. 880) like an appealing image and popularity. In fact, Dittmar et al. (2014) conclude that materialism may be best defined as a cluster of beliefs and values rather than a mere desire for money and material goods. In line with this finding, the present paper addresses materialism as a value an individual may endorse. Notably, almost all of us place at least some importance on materialistic aims, but for some materialism is at the centre of their value system (Kasser, 2002; Kasser, 2016).

Kasser, Ryan, Couchman, and Sheldon (2004) propose two main pathways to explain how people become inclined in materialistic values. First, people may become materialistic because of experiences that induce insecurities (Kasser et al., 2004). Naturally, people are motivated to engage in behaviours that fulfil their needs to feel safe, autonomous, competent, and connected to others for psychological growth to occur (Maslow, 1954; Ryan & Deci, 2000). People feel insecure, when they are exposed to environments and experiences that frustrate or hinder the satisfaction of these basic psychological needs. Thus, Kasser et al. (2004) suggest that experiences undermining the satisfaction of psychological needs can cause individuals to endorse materialistic values as an attempt to compensate for distressing feelings of insecurity, like doubts about self-worth, effective problem-solving, or safety. The second pathway suggests that people become materialistic because of exposure to social models that encourage materialistic values (Kasser et al., 2004). From birth onwards, individuals are exposed to messages favouring the significance of money and possessions delivered through parental values, the lifestyles of family members and peers, and messages found in popular culture, such as in the media. Therefore, Kasser et al. (2004) suggest that materialistic models and values exert direct influences on the development of materialistic value orientations through socialisation, internalisation (Banerjee & Dittmar, 2008), and modelling (Bandura, 1971). People tend to adopt to these materialistic models and thus, strive to attain materialistic goals as well. These two pathways involved in the development of materialistic value orientations might as well interact so that people experiencing higher levels of insecurity may be more sensitive to the impact of messages supporting acquisitiveness, which may in turn make them feel even more insecure, and vice versa (Kasser et al., 2004). This finds support by Richins (1995) who points out that, for instance, advertisements induce upward social comparisons which make viewers feel inferior resulting in reinforcing and maintaining feelings of insecurity underlying materialistic individuals. Further, individuals with materialistic value orientations are more concerned with social comparison (Kim, Callan, Gheorghiu, & Matthews, 2017), all of which make materialistic individuals more likely to be attentive to materialistic messages.

Since we now know how people become materialistic, what are the outcomes of these materialistic pursuits? A substantial body of evidence drawing from correlational as well as experimental studies demonstrates that a number of experiences, feelings, and behaviours of materialistic individuals are associated with a diminished quality of life. For example, compared to individuals who are low in materialistic values, individuals who prioritise materialistic values have worse money management skills (Donnelly, Ksendzova, & Howell, 2013), and create more debt (Richins, 2011). Kasser and Ryan (1993) have demonstrated that people with materialistic values focus more on obtaining rewards than on the inherent fun of activities. Vansteenkiste, Simons, Lens, Sheldon, and Deci (2004) have shown that experimentally framing tasks in terms of extrinsic (rather than intrinsic) goals resulted in decreased depth of processing, persistence, and performance in high school and college students. Materialism also fosters social isolation. People with materialistic values have relatively short relationships that are characterised by emotional extremes and conflict rather than on trust and happiness (Kasser & Ryan, 2001). Further, materialistic individuals place less significance on values such as affiliation (Kasser & Ryan, 1993), show less gratitude (Kashdan & Breen, 2007), and empathy (Sheldon & Kasser, 1995). Finally, materialistic individuals hold more negative attitudes about the environment and engage in less positive and more negative behaviours related to the environment (Hurst, Dittmar, Bond, & Kasser, 2013).

Most importantly, however, are materialistic individuals happier? A substantial body of evidence including correlational and experimental studies has proven materialistic values to be associated with diminished personal well-being across a broad array of well-being constructs. Individuals holding materialistic values consistently report lower subjective well-being (for a review see Kasser, 2002) and this across many different countries worldwide (e.g. Kim, Kasser, & Lee, 2003; Schmuck, Kasser, & Ryan, 2000). Findings of the latest meta-analysis to date investigating the relationship between individuals' materialistic orientation and their personal well-being support the evidence reviewed thus far. Dittmar et al. (2014) find that materialism correlates significantly and negatively with well-being suggesting that the more strongly individuals hold materialistic values, the poorer their personal well-being.

Minimalist life orientations as a tool for happiness

As we have seen thus far, individuals in Western societies strive for happiness just as much as they strive for wealth and possessions (e.g. Diener, 2000; Kasser, 2002). Moreover, people tend to believe that they would be happier if they were richer and thus that money can buy happiness (Kahneman et al., 2006). However, a substantial body of evidence indicates that money or rather holding wealth, popularity, and physical appearance as relatively important degrades an individual's personal well-being and psychological health. As a matter of fact, "studies document that strong materialistic values are associated with a pervasive undermining of people's well-being, from low life satisfaction and happiness, to depression and anxiety, to physical problems such as headaches, and to personality disorders, narcissism, and antisocial behavior" (Kasser, 2002, p. 22). Noteworthy, materialistic values also negatively affect health and happiness of many others as less empathy and intimacy are present in relationships of people with materialistic values (e.g. Sheldon & Kasser, 1995). They may even lead to further destructive consequences for the broader community when those in power objectify others in their pursuit of wealth and status (Kasser, 2002). Further, materialistic values lead to destructive consequences for Earth as they increase our vulnerability to serious social and environmental problems (Hurst et al., 2013) while impairing human cooperation and solution-finding processes to work against these problems (Kasser et al., 2004).

Considering this robust negative association between materialism and an individual's personal well-being, a logical consequence for people would be to focus less on materialistic pursuits and more on endeavours that actually do lead to happiness and satisfaction. Importantly, research shows that SWB is malleable and that the choices people make in life can influence their long-term SWB. Therefore, choosing a contrary lifestyle that holds contrary values should improve an individual's personal well-being. Schwartz's (1992) cross-cultural work shows that values are organised in a circumplex such that every value relates with some values and conflicts with other values. Materialistic values stand in conflict with intrinsic values for community, affiliation, self-acceptance goals, and spirituality (Grouzet et al., 2005). In line with this, research has shown that intrinsic values are associated with better personal well-being as people favouring intrinsic values report enhanced happiness, greater psychological health, better interpersonal relationships, more contribution to the community, and more concern for ecological issues (Kasser & Ryan, 1993, 2001; Sheldon & Kasser, 1995; Sheldon & McGregor, 2000). Thus, how could individuals in contemporary culture live more fulfilling and meaningful lives without centring on materialistic values? Could the minimalist lifestyle concept or minimalism be an alternative?

Minimalism

The minimalist lifestyle or minimalism can be thought of as one specific form within the broad construct of Voluntary Simplicity (VS). VS is often viewed as a general term for various forms of non-materialistic lifestyles that reject high consumption (Kala, Galčanová, & Pelikán, 2017). "Voluntary" refers to an intentional choice based on the realisation or critical awareness that society is out of control regarding its overconsumption and consequent sparseness of resources resulting in a value-driven conversion. "Simplicity" refers to the attenuation of involuntary and unnecessary daily complications. Thus, simplifiers might voluntarily choose to consume sustainably, to recycle and reuse, or to prefer smaller scale forms of living (Vannini & Taggart, 2013). Importantly, however, voluntary simplicity is distinct of poverty as poverty is not a deliberate choice ("involuntary simplicity"). Voluntary Simplicity focuses on the reduction of material consumption and removal of clutter from an individual's life without feelings of deprivation (Ballantine & Creery, 2010) "in order to free one's resources, primarily money and time, to seek satisfaction through non-material aspects of life" (Huneke, 2005, p. 528). Minimalism can be found within the broad diversity of Voluntary Simplicity lifestyles. Minimalism as a lifestyle is about a "reassessment of your priorities so that you can strip away the excess stuff – the possessions and ideas and relationships and activities – that don't bring value to your life" (Wright, 2010). Thus, the reduction of physical possessions is often a result of minimalism, but not minimalism itself (Wright, 2010). Fields Millburn and Nicodemus (2016, p. 24) ironically state that

> to be a minimalist you must live with fewer than 100 things, and you can't own a car or a home or a TV, and you can't have a career, [...] you can't have children, and you have to be a young white male from a privileged background.

Although this could be true, in most cases it is not and rather represent misconceptions of what minimalism supposedly is about. In reality, minimalism is a tool that helps to find freedom from "the trappings of the consumer culture we've built our lives around" and thus can be used to live a meaningful life (Fields Millburn & Nicodemus, n.d.). However, there is nothing inherently wrong about owning material possessions (Fields Millburn & Nicodemus, n.d.). But, we tend to assign too much meaning to material possessions instead of health, relationships, passions, personal growth, and contribution (Fields Millburn & Nicodemus, n.d.). Minimalism as a voluntary lifestyle choice, on the other hand, allows an individual to make conscious and deliberate decisions about what is important in an individual's life to make it meaningful. And as one frees up resources while reassessing one's priorities, these new resources can be used for life domains one actually does value (Wright, 2010). How exactly minimalism is translated in one's everyday life can differ drastically from one person to another as there is no one set of rules or a handbook to follow. What each and every one considers valuable in one's life and what non-essential is subjective (Fields Millburn & Nicodemus, n.d.). Even though individuals may embrace minimalism differently, each path leads to the same benefits. Leo Babauta (n.d.) from mnmlist.com describes benefits such as less stress, less debt, more time for love, peace and joy. Likewise, "The Minimalists" claim that minimalism has helped them to grow as individuals, to pursue their passions, to experience real freedom, to discover purpose in their lives, and, most importantly, to find lasting happiness. They continue that minimalists "search for happiness not through things, but through life itself". In sum, "minimalism is a tool to rid yourself of life's excess in favor of focusing on what's important—so you can find happiness, fulfillment, and freedom" (Fields Millburn & Nicodemus, n.d.).

Thus far, it seems that the present consumer culture is not set up for alternative lifestyles that do not value materialism. Hence, minimalists need to overcome various social constraints and the decision as well as the path to becoming a minimalist seem to require extended problem-solving processes since novice minimalists are unfamiliar with alternatives. In line with Dialogical Self Theory I argue that these internal inconsistencies are represented by multiple positions within the self and this emerging multiplicity of the minimalist's self results in a constant dialogue demanding synthesis.

From the perspective of the dialogical self theory

Individuals following the minimalist lifestyle claim to live meaningful and happy lives since their transition towards this niche lifestyle. Initial findings indeed suggest that simpler living is associated with greater happiness and life satisfaction (Alexander & Ussher, 2012; Boujbel & D'Astous, 2012) which provides further support that minimalism is a viable and desirable alternative to higher consumption lifestyles. Correspondingly, in recent years research interest has slowly shifted towards Voluntary Simplicity (e.g. Etzioni, 1998; Huneke, 2005) and the study of motivations, practices, and individual characteristics of simplifiers. However, thus far research has neglected to examine the decision-making processes underlying the transition towards becoming a minimalist. The increasing online media attention given to the practice of Minimalism is thought to reflect the number of people now adopting this lifestyle (Ballantine & Creery, 2010) and highlights that further research is required to better understand how individuals decide to adopt this alternative lifestyle. In the present paper, I propose to use Dialogical Self Theory (DST) as a general framework to explain the decision-making processes involved in the transition from a materialist to a minimalist.

Hermans, Kempen, and Van Loon (1992) conceptualise the self as multiple, fluctuating, and relatively autonomous I-positions through which the self continuously moves while adapting to changes in situation and time. Imaginatively, the individual endows each position with a "voice", and thereby establishes dialogical relations between different (and opposing) I-positions. These voices interact like "characters in a story" (Hermans et al., 1992, p. 28) resulting in a complex, narratively structured self with a hierarchy of positions (Hermans, 1996; Hermans, 2001). Hermans et al. (1992) further elaborate that over time, the self assumes different positions, as voices are influenced by modes of externalisation, reflecting the internal discussions within a person's mind and their ongoing interaction with the society and culture at large. Moreover, as positions move within the hierarchy in response to change, a predominant position can quickly become suppressed, while a previously neglected position moves to the forefront (Hermans, 2004, 2018; Hermans & Hermans-Konopka, 2010). Therefore, DST can be used to attend to multiple I-positions and the dialogical relations between positions to understand how internal inconsistencies are resolved. To further elaborate on my proposition the following case illustration depicts how DST explains the decision-making processes involved in the transition from a materialist to a minimalist.

Case illustration: dialogical self in the minimalists

Since the self or an individual's identity continuously adapts to changes in situation and time (Hermans et al., 1992) by moving through its multiple I-positions, specific situations or times trigger specific I-positions which in turn impact the individual's identity construction. Since values are organised in a circumplex fashion (Schwartz, 1992), almost all of us place at least some importance on materialistic aims as well as on intrinsic/minimalistic aims, but for some, one of each is at the centre of their values system (Kasser, 2016). Thus, the materialistic I is part of the self just like the minimalistic I. However, I argue that over time more and more situations trigger the minimalistic I instead of the materialistic I.

How does this apply to Joshua Fields Millburn and Ryan Nicodemus? These two authors call themselves "The Minimalists" and their transition from being materialists to becoming minimalists will serve to illustrate the steps of the process of identity change. In 1998, during Fields Millburn's and Nicodemus' last year of high school, the two best friends concluded that the key to happiness must be money. Both of them grew up in poor, dysfunctional households with unhappy parents which made them decide that passing an arbitrary threshold of $50,000 a year would somehow lead to their happiness. Thus, "The Minimalists" both started to work after their graduation and focused on earning money and status ever since. In 2009, more than 10 years later, "The Minimalists" each lived their version of the American Dream with high paying jobs (more than $50,000 a year!). However, they also still felt discontent. Thus, they tried to purchase happiness by buying "everything [...] consumer culture told us would make us happy" (p. 9). But their "abundance of stuff" (p. 3) didn't make them happy, they only felt more discontent. Notably, up to this point their lives comply with the two main pathways to becoming materialistic (Kasser et al., 2004). Their basic psychological needs were undermined which caused them to endorse materialistic values as an attempt to compensate for distressing feelings of insecurity, and the influences of materialistic models like consumer culture or their family and friends led them to develop materialistic values as well. In line with the empirical evidence on the pitfalls of materialism, Fields Millburn and Nicodemus had all the issues described.

They had poor money management skills (Donnelly et al., 2013), they created debt (Richins, 2011), they focused more on obtaining rewards (Kasser & Ryan, 1993), and they were unhappy (Kasser, 2002). It seems like both Fields Millburn's and Nicodemus' materialistic I can be thought to be a core position (Hermans & Hermans-Konopka, 2010) and to be at the peak of their hierarchy of positions at this time. In late 2009, Fields Millburn's mother died. This critical life event made him question every aspect of his life since he came to the realisation that "the relentless pursuit of riches doesn't lead to a meaningful life" (p. 16). It was about this time that Fields Millburn and Nicodemus had another conversation about happiness with the conclusion that they want to find out what was making them unhappy. This change in situation and time triggered a thus far neglected position, the minimalistic I, which can be thought of as a counter-position to the materialistic I, and as being at the very bottom within the hierarchy of positions.

From that point onwards, these two opposing voices negotiate their positions to resolve inconsistencies since "The Minimalists" "decided to take an inventory of [their] lives" (p. 17). They began with identifying their anchors which can be anything that made them feel stuck and kept them from growing and thus, prevented them from living happy and fulfilled lives. In other words, their materialistic I got raised to question by the minimalistic I. Fields Millburn and Nicodemus each made a list on which they wrote down anything that might be an anchor such as certain relationships with people that didn't add value to their lives, bills and debts, household clutter and other small things that needed their time and attention. The opposing voices of the materialistic and the minimalistic I had to negotiate regarding each identified anchor. The minimalistic voice won over its counter-position with each anchor that got written down on the list as well as with each anchor "The Minimalists" got disposed of, as depicted in the following example (p. 16):

> We had discovered "getting what we wanted" (large houses, bigger paychecks, material possessions, and corporate awards) wasn't making us happy, so we wanted to identify what was anchoring us […]. We decided that ridding ourselves of many of these anchors […] would let us reclaim much of our own time, which could then be spent in more meaningful ways. […] For example, every extra penny Joshua earned was spent on making extra payments towards his debts. No more trips, vacations, fancy dinners […].Other major anchors were addressed in a similar fashion. We eventually jettisoned many of our possessions, eliminating the excess in favor of things we liked and enjoyed – the things we actually used in our daily lives. Over the course of two years, our anchors of old age were no longer weighing us down.

Thus, during this reassessment of priorities, the materialistic I gets corrected while the voice of the minimalistic I gets louder since "The Minimalists" converse their values from materialistic to intrinsic. Likewise, the position of these two voices in the hierarchy of their self shift, and both can now be found on the same hierarchy level. This means that although the materialistic I is not a core position anymore and is not as dominant as it used to be, the minimalistic I is also not yet strengthened enough to fully overpower the materialistic I.

By disposing of each of these anchors over a period of two years, "The Minimalists" reclaimed time which they could then spent in more meaningful ways. However, they discovered what was most important to them through "trial and error" (p. 30). Throughout their "inventory of their lives" there might have been situations in which the self momentarily moved more in the direction of the materialistic I than the minimalistic I. One of these "setbacks" is described in the following (pp.19):

> Joshua was faced with the dilemma of what to do with his mother's stuff after her death – what to do with those sentimental items we tend to hold on to in perpetuity? […] Joshua did what any son would do: He rented a large U-Haul truck. Then he called a storage facility back in Ohio to make sure they had a big enough storage unit. […] At first Joshua didn't want to let go of anything. […] So instead of letting go, he planned to cram every trinket and figurine and piece of oversized furniture into that tiny storage locker in Ohio. [Then] he realized his retention efforts were futile. He could hold on to her memories without her stuff [and] didn't need a storage locker filled with her belongings to remember her. So Joshua called U-Haul and canceled the truck.

Since the materialistic I used to be a core position and the minimalistic I aims to become a core position within the self, these two opposing voices negotiate not only with each other, but beyond that also with further I-positions as the functioning of these I-positions depend on the core position. For example, Joshua is also the son of a recently deceased mother, a husband, or a member of the American culture. All these internal and collective voices discuss how the desired life reorientation can be arranged with some voices being closer to the minimalistic I than others. Consequently, there might be setbacks since the minimalistic voice cannot always correct or complement all the other voices. However, "The Minimalists" stayed persistent in their journey towards a more meaningful and happy life and through their experimentation they discovered that five values allow them to live a meaningful life: Health, relationships, passions, growth, and contribution. These "Five Values are the areas [they] changed in [their] lives that had the largest positive effect and resulted in more satisfaction and contentment" (p. 30). Notably, their "Five Values" resemble the five ways to well-being suggested by the New Economics Foundation (NEF, 2011) which are connecting with others, taking notice of one's environment, to keep learning, to be active, and doing favours for others. Further, these "Five Values" fall within the cluster of intrinsic values. Now, the selves of "The Minimalists" moved back to the minimalistic I as it stood up to all the other voices and most importantly to the materialistic I in the long run. The hierarchy positions of these two voices have exchanged: the minimalistic I has become more dominant than the materialistic I.

While getting rid of their anchors they also searched for examples of people who went through a similar process. Like this, "The Minimalists" stumbled upon the concept of minimalism. They found that minimalism is for "anyone interested in living a simpler, more intentional life [… and] for anyone who want[s] to focus on the important aspects in life, rather than the material possessions so heavily linked to success and happiness by our culture" (p. 24). Thus, "The Minimalists" fully embraced the concept of minimalism. The minimalistic I has become the core position and has completely been incorporated in the identity of "The Minimalists" which completed their identity transformation from a materialist to a minimalist. Fields Milburn's and Nicodemus' beliefs and values changed from placing importance on wealth, possessions, and status to placing importance on health, relationships, passions, growth, and contribution. Since values are organised in a circumplex (Schwartz, 1992), the materialistic I will remain an I-position of "The Minimalists". Hence, as various I-positions within the self engage in dialogue, they constantly cultivate themselves – reconstructing, renegotiating and reorganising meanings of selves, in relationship to each other and the environment. Thus, there will remain self-making as long as there is dialogue (Valsiner, 1999).

This case illustration suggests five stages to be involved in the identity transformation from a materialist to a minimalist (see Table 1). The first stage is about finding discontent in one's life and can be seen as a prerequisite to realising that something needs to be changed in one's life (Stage 2). This decision marks the beginning of a reassessment of priorities (Stage 3). The following introspection forces the individual to confront himself with who he is and with what is important in his life to find out how he can make his life more fulfilling and meaningful. In stage four the resulting reorientation solidifies in all desired life domains which completes the identity transformation. Now, intrinsic values are at the centre of an individual's value system (Stage 5) which gives an individual the potential to foster a happy and meaningful life. These five stages emphasise again that minimalism needs to be understood as a tool and not as a means in itself and can be thought of as a form of self-

Table 1. Stages involved in the transition from materialist to minimalist.

	Materialist to minimalist	Dialogical self
Stage 1	Finding discontent in one's life	Materialistic I as core position
Stage 2	Realisation of and decision for change	Materialistic I as core position, minimalistic I gets triggered as counter position
Stage 3	Beginning of reassessment (introspection)	Both I-positions negotiate to resolve inconsistencies and can now be found at the same hierarchy level
Stage 4	Solidification	Exchange of hierarchy positions with minimalistic I now being superior to the materialistic I
Stage 5	Final identity transformation	Minimalistic I has become core position, full integration in the identity, materialistic I remains I-position at the bottom of the hierarchy of the self

guidance to reach self-development or self-optimisation. By engaging in internal and external dialogues the self guides and counsels itself in its self-making. Therefore, DST has proven to be a useful framework to understand the identity transformation from a materialist to a minimalist in my chosen example.

Conclusion

In the present paper I proposed to use Dialogical Self Theory (Hermans et al., 1992) as a general framework to explain the identity and value transformation from a materialist to a minimalist.

Since subjective well-being is malleable (Tay & Kuykendall, 2013), the choices people make in life can influence their long-term SWB. Hence, minimalism, as one form of alternative lifestyles, raises the possibility that individuals in contemporary consumer culture could live more fulfilling and meaningful lives by reducing their consumption and focusing on intrinsic values. However, thus far research has neglected to examine the decision-making processes underlying the transition towards becoming a minimalist. This exploratory attempt to use DST as a framework to better understand the decision-making processes involved in the transition from a materialist to a minimalist offers initial insights in the different stages involved in this identity transformation. The exemplary transition of Joshua Fields Millbrun and Ryan Nicodemus ("The Minimalists") suggests five stages: finding discontent, decision for change, reassessment, solidification, final identity transformation. Their identity transformations result out of the dialogical relations between an individual's multiple I-positions and primarily between the materialistic and the minimalistic I. Thus, this transitional process disguised as minimalism can be thought of as a form of self-guidance to reach self-development or self-optimisation. By engaging in internal and external dialogues the self guides and counsels itself in its self-making. Thus, the overall transformation or rather minimalism can also be seen as taking a broader position, namely a meta-position from which all involved I-positions are considered (Hermans & Hermans-Konopka, 2010). Hermans and Hermans-Konopka (2010, p. 147) explain that a meta-position permits a certain distance from other positions and hence, "facilitates the creation of a dialogical space (in contact with others or with oneself) in which positions and counter-positions engage in dialogical relationships". This, in turn, allows the self a broader perspective for decision-making and to find one's direction in life just like the principles of minimalist lifestyle posit. Therefore, DST has proven to be a useful framework to understand the identity transformation from a materialist to a minimalist in my chosen example. In particularly, DST provides an alternative perspective in which conflicts and contradictions are not necessarily negative, but they offer the potential for innovations in the self. Further, the emerged knowledge may in turn help to derive and to develop means which potentially support and facilitate an individual's transition. Moreover, it may help to understand challenges individuals face which is important for those interested in spreading minimalism to mainstream society (e.g. policies interested in promoting happy and sustainable living; McGouran & Prothero, 2016). Initial insights in challenges individuals face guide means to overcome barriers to simplification in a culture that values materialism. Considering the detrimental consequences of materialistic values on personal well-being, community, society, and Earth (see Kasser, 2002) it seems inevitable to find ways to encourage individuals to transition to alternative, materially simpler lifestyles in the near future. Moreover, researchers generally agree (e.g. Alexander & Ussher, 2012; Etzioni, 1998) that any form of materially simple lifestyles is a necessary and effective response to the ecological crisis and are the requirement to transition to a sustainable and just society since overconsumption in affluent societies is considered as the main contributor of many world problems. Thus, engaging in an alternative, non-materialistic and anti-consumption lifestyle seems to have many impactful benefits for the society and the individual beyond increased personal well-being.

However, I exemplary took a look at the identity transformation of two minimalists. Can the same transition including its five stages be described for any individual becoming a minimalist or are there differences to note? What might these differences depend on? Do all minimalists experience a triggering situation resulting in the minimalistic I as a counter position to the materialistic I and is it possible that some individuals are not even aware of their gradual identity transformation? Future studies should address these questions by examining the transition of several minimalists to further refine the proposed stages involved in this identity transformation.

To conclude, the present paper was an exploratory commentary proposing Hermans' Dialogical Self Theory as a useful framework to explain the decision-making processes involved in the transition from materialist to minimalist. Considering that every life domain (including family and friends) influencing an individual's self is re-evaluated to assess its value to oneself and considering the social constraints put on living simpler lives, it seems that an identity transformation from being a materialist to becoming a minimalist is a process that requires high individualistic motivation and perseverance to change one's lifestyle. However, DST as a framework supports the fact that happiness can be fostered out of a thorough introspection irrespective of social challenges simplifiers might face. Therefore, dialogical relations of different I-positions created by the meta-position of minimalism serve as a tool, as a form of self-guidance to develop or transform one's identity. DST as a framework for my proposition thus evokes and stimulates further research in the increasingly important field of alternative lifestyles.

Acknowledgement

I would like to thank Jaan Valsiner, Hubert Hermans, Frans Meijers as well as the two anonymous referees for their careful reading of my manuscript and their insightful comments and suggestions.

Disclosure statement

No potential conflict of interest was reported by the author.

References

Alexander, S., & Ussher, S. (2012). The voluntary simplicity movement: A multi-national survey analysis in theoretical context. *Journal of Consumer Culture, 12*(1), 66–86.

Babauta, L. (n.d.). mnmlist: minimalist FAQs. [Blog post]. Retrieved from http://mnmlist.com/minimalist-faqs/

Ballantine, P. W., & Creery, S. (2010). The consumption and disposition behaviour of voluntary simplifiers. *Journal of Consumer Behaviour, 9*, 45–56.

Bandura, A. (1971). *Social learning theory*. Morristown, NJ: General Learning Press.

Banerjee, R., & Dittmar, H. (2008). Individual differences in children's materialism: The role of peer relations. *Personality and Social Psychology Bulletin, 34*(1), 17–31.

Berry, D. S., & Hansen, J. S. (1996). Positive affect, negative affect, and social interaction. *Journal of Personality and Social Psychology, 71*, 796–809.

Boujbel, L., & D'Astous, A. (2012). Voluntary simplicity and life satisfaction: Exploring the mediating role of consumption desires. *Journal of Consumer Behaviour, 11*, 487–494.

Busseri, M. A., & Sadava, S. W. (2011). A review of the tripartite structure of subjective well-being: Implications for conceptualization, operationalization, analysis, and synthesis. *Personality and Social Psychology Review, 15*, 290–314.

Diener, E. (1984). Subjective well-being. *Psychological Bulletin, 95*, 542–575.

Diener, E. (2000). Subjective well-being: The science of happiness and a proposal for a national index. *American Psychologist, 55*(1), 34–43.

Diener, E., & Chan, M. (2011). Happy people live longer: Subjective well-being contributes to health and longevity. *Applied Psychology: Health and Well-Being, 3*, 1–43.

Diener, E., Heintzelman, S. J., Kushley, K., Tay, L., Wirtz, D., Lutes, L. D., & Oishi, S. (2017). Findings all psychologists should know from the new science on subjective well-being. *Canadian Psychology/Psychologie Canadienne, 58*(2), 87–104.

Diener, E., & Oishi, S. (2000). Money and happiness: Income and subjective well- being across nations. In E. Diener, & E. M. Suh (Eds.), *Culture and subjective well- being* (pp. 185–218). Cambridge, MA: MIT Press.

Diener, E., Oishi, S., & Tay, L. (2018). Advances in subjective well-being research. *Nature Human Behaviour, 2*, 253–260. doi:10.1038/s41562-018-0307-6

Dittmar, H. (2008). *Consumer culture, identity and well-being: The search for the "good life" and the "body perfect"*. Hove, England: Psychology Press.

Dittmar, H., Bond, R., Hurst, M., & Kasser, T. (2014). The relationship between materialism and personal well-being: A meta-analysis. *Journal of Personality and Social Psychology, 107*(5), 879–924.

Donnelly, G., Ksendzova, M., & Howell, R. T. (2013). Sadness, identity, and plastic in over-shopping: The interplay of materialism, poor credit management, and emotional buying motives in predicting compulsive consumption. *Journal of Economic Psychology, 39*, 113–125.

Etzioni, A. (1998). Voluntary simplicity: Characterization, select psychological implications, and societal consequences. *Journal of Economic Psychology, 19*(5), 619–643.

Fields Millburn, J., & Nicodemus, R. (2016). *Minimalism – live a meaningful life*. Missoula, Montana: Asymmetrical Press.

Fields Millburn, J., & Nicodemus, R. (n.d.). What is Minimalism? [Blog post]. Retrieved from https://www.theminimalists.com/minimalism/

Grouzet, F. M. E., Kasser, T., Ahuvia, A., Dols, J. M. F., Kim, Y., Lau, S., … Sheldon, K. M. (2005). The structure of goal contents across 15 cultures. *Journal of Personality and Social Psychology, 89*, 800–816.

Hermans, H. J. M. (1996). Voicing the self: From information processing to dialogical interchange. *Psychological Bulletin, 119*, 31–50.

Hermans, H. J. M. (2001). The dialogical self: Toward a theory of personal and cultural positioning. *Culture & Psychology, 7*(3), 243–281.

Hermans, H. J. M. (2004). Introduction: The dialogical self in a global and digital age. *Identity: An International Journal of Theory and Research, 4*(4), 297–320.

Hermans, H. J. M. (2018). *Society IN the self: A theory of identity in democracy*. New York: Oxford University Press.

Hermans, H. J. M., & Hermans-Konopka, A. (2010). *Dialogical self theory: Positioning and counter-positioning in a globalizing society*. Cambridge: University Press.

Hermans, H. J. M., Kempen, H. J. G., & Van Loon, R. J. P. (1992). The dialogical self: Beyond individualism and rationalism. *American Psychologist, 47*, 23–33.

Huneke, M. E. (2005). The face of the un-consumer: An empirical examination of the practice of voluntary simplicity in the United States. *Psychology & Marketing, 22*(7), 527–550.

Hurst, M., Dittmar, H., Bond, R., & Kasser, T. (2013). The relationship between materialistic values and environmental attitudes and behaviors: A meta-analysis. *Journal of Environmental Psychology, 36*, 257–269.

Kahneman, D., & Deaton, A. (2010). High income improves evaluation of life but not emotional well-being. *Proceedings of the National Academy of Sciences, 107*(38), 16489–16493.

Kahneman, D., Krueger, A. B., Schkade, D., Schwarz, N., & Stone, A. A. (2006). Would you be happier if you were richer? A focusing illusion. *Science, 312*, 1908–1910.

Kala, L., Galčanová, L., & Pelikán, V. (2017). Narratives and practices of voluntary simplicity in the Czech post-socialist context. *Czech Sociological Review, 53*(6), 833–855.

Kashdan, T. B., & Breen, W. E. (2007). Materialism and diminished well-being: Experiential avoidance as a mediating mechanism. *Journal of Social and Clinical Psychology, 26*, 521–539.

Kasser, T. (2002). *The high price of materialism*. Cambridge: A Bradford Book, The MIT Press.

Kasser, T. (2016). Materialistic values and goals. *Annual Review of Psychology, 67*, 489–514.

Kasser, T., & Ryan, R. M. (1993). A dark side of the American dream: Correlates of financial success as a central life aspiration. *Journal of Personality and Social Psychology, 65*(2), 410–422.

Kasser, T., & Ryan, R. M. (2001). Be careful what you wish for: Optimal functioning and the relative attainment of intrinsic and extrinsic goals. In P. Schmuck, & K. M. Sheldon (Eds.), *Life goals and well-being: Towards a positive psychology of human striving* (pp. 116–131). Goettingen: Hogrefe & Huber.

Kasser, T., Ryan, R. M., Couchman, C. E., & Sheldon, K. M. (2004). Materialistic values: Their causes and consequences. In T. Kasser, & A. D. Kanner (Eds.), *Psychology and consumer culture: The struggle for a good life in a materialistic world* (pp. 11–28). Washington: American Psychological Association.

Kim, H., Callan, M. J, Gheorghiu, A. I., & Matthews, W. J. (2017). Social comparison, personal relative deprivation, and materialism. *British Journal of Social Psychology, 56*, 373–392.

Kim, Y., Kasser, T., & Lee, H. (2003). Self-concept, aspirations, and well-being in South Korea and the United States. *The Journal of Social Psychology, 143*(3), 277–290.

Lount, R. B., Jr. (2010). The impact of positive mood on trust in interpersonal and intergroup interactions. *Journal of Personality and Social Psychology, 98*, 420–433.

Lyubomirsky, S., King, L. A., & Diener, E. (2005). The benefits of frequent positive affect: Does happiness lead to success? *Psychological Bulletin, 131*, 803–855.

Maslow, A. H. (1954). *Motivation and personality*. New York: Harper and Row.

McGouran, C., & Prothero, A. (2016). Enacted voluntary simplicity – exploring the consequences of requesting consumers to intentionally consume less. *European Journal of Marketing, 50*(1/2), 189–212.

NEF. (2011). *Five ways to wellbeing: New application, new ways of thinking.* London: New Economics Foundation. Retrieved from http://neweconomics.org/2008/10/five-ways-to-wellbeing-the-evidence/

Oishi, S., Diener, E., & Lucas, R. E. (2007). The optimal levels of well-being: Can people be too happy? *Perspectives on Psychological Science, 2*, 346–360.

Richins, M. L. (1995). Social comparison, advertising, and consumer discontent. *American Behavioral Scientist, 38*, 593–607.

Richins, M. L. (2011). Materialism, transformation expectations, and spending: Implications for credit use. *Journal of Public Policy and Marketing, 30*, 141–156.

Ryan, R., & Deci, E. (2000). Self-determination theory and the facilitation of intrinsic motivation, social development, and well-being. *American Psychologist, 55*(1), 68–78.

Schmuck, P., Kasser, T., & Ryan, R. M. (2000). Intrinsic and extrinsic goals: Their structure and relationship to well-being in German and U.S. College students. *Social Indicators Research, 50*, 225–241.

Schwartz, S. H. (1992). Universals in the content and structure of values: Theory and empirical tests in 20 countries. In M. Zanna (Ed.), *Advances in experimental social psychology* (Vol. 25, pp. 1–65). New York: Academic Press.

Sheldon, K. M., & Kasser, T. (1995). Coherence and congruence: Two aspects of personality integration. *Journal of Personality and Social Psychology, 68*, 531–543.

Sheldon, K. M., & McGregor, H. (2000). Extrinsic value orientation and the tragedy of the commons. *Journal of Personality, 68*, 383–411.

Tay, L., & Kuykendall, L. (2013). Promoting happiness: The malleability of individual and societal subjective wellbeing. *International Journal of Psychology, 48*, 159–176.

Twenge, J. M., Freeman, E. C., & Campbell, W. C. (2012). Generational differences in young adults' life goals, concern for others, and civic orientation, 1966–2009. *Journal of Personality and Social Psychology, 102*(5), 1045–1062.

Twenge, J. M., Gentile, B., DeWall, C. N., Ma, D., Lacefield, K., & Schurtz, D. R. (2010). Birth cohort increases in psychopathology among young Americans, 1938–2007: A cross-temporal meta-analysis of the MMPI. *Clinical Psychology Review, 30*, 145–154.

Valsiner, J. (1999). I create you to control me: A glimpse into basic processes of semiotic mediation. *Human Development, 42*, 26–30.

Vannini, P., & Taggart, J. (2013). Voluntary simplicity, involuntary complexities, and the pull of remove: The radical ruralities of off-grid lifestyles. *Environment and Planning, 45*, 295–311.

Vansteenkiste, M., Simons, J., Lens, W., Sheldon, K. M., & Deci, E. L. (2004). Motivating learning, performance and persistence: The synergistic effects of intrinsic goal contents and autonomy-supportive contexts. *Journal of Personality and Social Psychology, 87*, 246–260.

Whelan, D. C., & Zelenski, J. M. (2012). Experimental evidence that positive moods cause sociability. *Social Psychological and Personality Science, 3*, 430–437.

Wright, C. (2010, September 15). Minimalism explained [Blog post]. Retrieved from http://exilelifestyle.com/minimalism-explained/

The search for inner silence as a source for *Eudemonia*

Olga V. Lehmann, Goran Kardum and Sven Hroar Klempe

ABSTRACT

In this paper we reflect upon inner silence as a path to embrace eudemonia in relation to Dialogical Self Theory. We suggest that the awareness of the dynamics of different I-positions can empower the person to feel more freedom over their thoughts, feelings and actions. Thus, inner silence can promote the capacity to activate meta-positions or de-positioning, which are shifts of attentional foci in the stream of consciousness. The practice of inner silence can also promote the experience of genuine dialogues, by encouraging the flow and flexibility of the self, and by strengthening the connection with the experience of values. We also suggest a number of implications for counselling associated with self-exploration and self-transcendence, such as meditation, contemplation and prayer.

The question of what happiness is has historically been one of the trickiest inquiries in psychology and other disciplines. For example, the old Greek term *eudaimonia*, and its different connotations, suggest several challenges for achieving an understanding of what "happiness" is. *Eudemonia* consists of two words: *eu* meaning "well" or "good" and *daimon* that can be translated with "spirit". However, modern associations with the notion of the "spirit" do not necessarily coincide with those of Greek philosophy. Socrates referred to the *daimon* when he said he was guided by an inner voice (Eriksen, 1972), and it is highly associated with the psyche. This implies that *daimon* can also be translated as "the power controlling the destiny of an individual" (Miller, 2015, p. 174). Such a definition raises the possibility of embracing happiness by finding a path towards freedom despite lifés constraints. Yet, how is one to strive for the values that promote onés ability to responsibly use his or her freedom of will? In this paper, we attempt to look at this problem through the lenses of silent experiences. Since the word "silence" suggests a wide diversity of experiences (Colum, 2011), we use the term "silence-phenomena" (Lehmann, 2018) to highlight the complexity of meanings and layers evoked by such a notion. We then proceed with a theoretical overview of the implications of silence-phenomena for eudemonic happiness in relation to Dialogical Self Theory, thereby raising several implications for counselling and work.

Silence-phenomena as a source for dialogue

Silence-phenomena can be understood by means of a meta-categorisation that involves silence, silences and silencing (Bruneau & Ishii, 1988; Orlandi, 1995). Silence refers to experiences of profound connection such as mystical and (or) aesthetic experiences, often described as the temporal dissolution of the perception of time and space. Silences involve social aspects of integration, such as turn-taking in communication or spaces that regulate different practices (e.g. waiting rooms in hospitals). Silencing reflects power dynamics, such as in the case of oppression or manipulation. Although these three notions are interconnected, we will mainly emphasise the experiences of "silence" as a source of genuine dialogues with others or with ourselves. According to Dialogical Self Theory, genuine dialogues diverge from conventional interactions between different I-positions. That is, genuine dialogues suggest an explorative and not judgmental attitude; one that encourages deep listening, honesty, respect for alterity, self-knowledge, as well as promoting innovation, such as finding win-win solutions during conflicts (Hermans & Hermans-Konopka, 2010a; Puchalska-Wasyl, 2010; Nir, 2016; van Loon & van den Berg, 2016). In addition, genuine dialogues encourage the kind of awareness for which silent experiences make room (Hermans & Hermans-Konopka, 2010a). Indeed, one of the key characteristics of silence-phenomena is that they foster attentional shifts (Lehmann, 2018). For example, silent experiences can promote an awareness of the contents of consciousness, such as the tensions and other dynamics among different I-positions (Lehmann & Valsiner, 2017).

Furthermore, silence-phenomena suggest an interdependence with language, sound, noises and (or) movements (Kurzon, 1998). In relation to language, silence-phenomena evoke signs that can be further analysed in psychology as part of the investigation of human experiences. However, silence-phenomena also illustrate the limited capacity of language to express the qualities of human experience, such as intense emotions and feelings (Lehmann, 2018). In this sense, the roles and effects of silence-phenomena in dialogue need to be explored further.

Inner silence and the dialogical self

In our everyday lives, silence-phenomena are breaks that open up our consciousness for the simultaneous coexistence of associations, affective arousals, and voices from diverse positionings of the self that are essential for the processes of meaning-making, decision-making and value-adding (Lehmann, 2018). According to Lazzari (2009), inner silences relate to the activation of the narrating self. That is, inner-silence promotes internal dialogues, not just a focus on one's own feelings, thoughts and behaviours. Thus, we understand the notion of inner silence to be a process that recalls two movements – one inward and one outward – in the dialogical landscape of the self. The movement inwards is one of self-exploration undertaken in order to focus on a dialogue with internal I-positions. The movement outwards recalls the process of engaging in a dialogue with others. These shifts of attention involve awareness of different I-positions present in the stream of consciousness, which in terms of Dialogical Self Theory is one possibility to develop the capacity of meta-positioning. A meta-position places one in the role of an observer of experience involving self-acceptance and insights into our experiences as human beings that have a strong affective basis (Hermans, 2003). Yet, our capacity to activate such a meta-positioning is a skill we need to train, and searching for inner silence can serve as a resource that can enable access to such an observer position.

In addition, inner silences are also related to the moments in which awareness is not fixed on particular I-positions, but rather transcends them or "de-positions" from them due to the intensity of feelings and emotions (Hermans & Hermans-Konopka, 2010b, p. 425). Lehmann (2018) identified silence as instants of profound connection with ourselves or with others, sometimes described as mystical, aesthetic, or dialogical encounters.

Thus, inner silence points at connections with spirituality. Some monastic traditions describe experiences of inner silence through the notion of stillness, which points towards the embracement of contemplation (Sarah & Diat, 2017). However, the idea of inner silence as a path for well-being can pave the way for an understanding of spirituality, but not necessarily. From either a secular or a religious perspective, one could say that inner silence is a source for establishing relational depth (Lehmann, 2018). Relational depth is a state of profound understanding, connection, and engagement both in our relationships with the Other (Mearns & Cooper, 2005) and with ourselves (Cooper, 2004). Thus, relational depth is a characteristic of authentic dialogues, or what Buber (1950) would call I-Thou encounters. These dialogues, in contrast to other forms of interaction, manifest themselves as dialectic, dynamic and innovative, fostering genuineness and transformation among their participants (Hermans & Hermans-Konopka, 2010b). Relational depth requires practice. Our stand, therefore, is that inner silence appears as a necessary condition for a person to become aware of different aspects of the inner life, such as the interaction and dialogue among I-positions. In addition, in order to achieve *eudaimonia*, it is crucial not just to remain within such realms of self-exploration, but also to respond accordingly to one's values.

Noise and other constraints on Eudemonic happiness

The World Health Organization (WHO) has recognised environmental noise as harmful pollution that threatens human health and has argued that it may be important for the public's health to update existing noise-related policies (Kim et al., 2012). Yet, the implications of noise and sound have also a metaphorical significance in terms of human experience as a dialogical realm. Humans are increasingly confronted by internal and external noise, which can limit our capacity to embrace inner silence. In terms of Dialogical Self Theory, this can be explained by difficulties in organising the repertoire of I-positions in the self, which leads to a cacophony in the mind (Hermans, personal communication, 9 May 2018). It is, therefore, necessary to explore alternatives that help to tune in to the dialogues between positionings of the self or with the Other that promote healthier organisations of mental processes. This is so, since threats to mental health such as depression are predicted to be by 2030 the primary disability in the world (WHO, 2017). Mindfulness, and other practices related to inner silence, has been increasingly recognised as promoting our capacity to immerse, detach, and move on from the identification with feelings and thoughts (David, 2016). Thus, developing habits that promote the search for inner silence and our faculty to direct our attention towards the realms of dialogue might favour our experience of eudemonic happiness and promote mental and spiritual health.

Yet, in order to experience genuine inner-silence, conducive external conditions (such as the absence of noise) or observations of one's thoughts through meta-positioning are not enough if one lacks the affective arousals that provide the motivation to appreciate the beauty in silence (among other motivations) (Tangene, 2017). Thus, rather than speaking of eudemonic happiness as a state one reaches, it can be better described as a "lifestyle characterised by the pursuit of virtue/ excellence, meaning/purpose, doing good/making a difference, and the resulting sense of fulfilment or flourishing" (Wong, 2011, p. 14). This is so, as *eudemonia* recalls man's search for meaning in relation to the experience of values, to the creation of valuable experiences for others or ourselves, and to our attitude towards impending suffering (Frankl, 1966/2000). In connection with this, relational depth, as an expression of our dialogical nature (Schmid, 2013), suggests movements between experiences of the Self and Otherness as a whole and the detachment that leads to focusing on particular I-positions. Both movements of connection and detachment within I-positioning and the person as a whole are necessary for the experience of eudemonic happiness. In practice, it is common to experience moments of disconnection that affect the quality of dialogues between people and discourage relational depth (Cooper, 2012). Thus, human beings long for genuine dialogues and for the permanence of moments of relational depth, while at the same time they face the uncertainty and tension caused by the realisation it is impossible to continue this endlessly (Lehmann, 2018).

Inner silence and verticality: insights from music

The different voices or layers in music relate to the simultaneous coexistence of instruments, voices and tones. Also, pauses and ruptures are distributed in different layers, and the polyphonic interaction between the layers forms the tensions in music. Taking this situation as a parallel to the complexity of the self, silence-phenomena recall the tension that the trajectories of I-positions imply (Lehmann & Valsiner, 2017). For example, silent settings can promote self-explorative processes where awareness of the activation of particular I-positions. Yet, engaging in practices of silence can also give an account of the patterns of thoughts and feelings related to such I-positions in the stream of consciousness, and to the tension that is generated in a process of stepping out of those patterns (Tangene, 2017; Lehmann, 2018). We argue that the multiplicity of I-positions and the interaction and (or) dialogue between them is a necessary condition for the pursuit of virtues and meaning in life; a pursuit that is characteristics of eudemonic happiness (Wong, 2011).

Furthermore, pauses in music raise the following question: What is happening now? Thus, the pause is not empty. It brings up all the possible outcomes the listener may come up with. An abundance of alternatives fills the pause. All the alternatives can only exist as long as the silence remains. In this sense, the pause is pregnant with much more than the music itself can contain. This learning from music reflects the fact that silence-phenomena bring in spaces where the potentiality of continuations, meanings and decisions are present for a short while (Lehmann, 2016). Taking music as a reference point, one could argue that the dialogicality of the self involves a composition of layers of tensions, where the different voices of the self-appear in different intensities across chronological time (Lehmann & Klempe, 2017). These tensions can represent the forms in which a human being experiences their own attempts in search for freedom despite one's constraints – a search that is a key aspect of eudemonic happiness (Miller, 2015).

The American composer John Cage expressed his vision of how silences are filled with an insurmountable abundance of sounds in his famous composition entitled, "4'33". This piece of "music" refers to 4 minutes and 33 seconds of "no-sounds". The purpose of the composition is to demonstrate that absolute silence is just an ideal or abstraction, and that this long-lasting pause, in fact, produces a lot of sounds as the audience starts to cough, scratch their legs, and otherwise move their bodies: "In fact, try as we may to make a silence, we cannot" (Cage, 1939/1961, p. 8). To be silent, therefore, is to open up to sounds of which we are normally not fully aware. As part of her doctoral research project, Lehmann (2018) created a silent time during an interdisciplinary course for Master's students in Norway. One of the activities of the class was to watch a musical video of a performance of "4'33". After analysing the diary entries of some of the students, the researcher found that silent settings acted as ruptures that made the students aware of different layers of tension, thereby evoking experiences of uncertainty and a contrast between expectations and reality. First, the long-lasting break was not experienced as mere emptiness. Second, the experience was filled up with something unexpected. Third, the tension between expectations and reality made the experience appear as undifferentiated, which also made the experience difficult to put into words. This third result indicates that the experience of this piece of music included an unspecified potentiality of meanings that could go in many different directions. Thus, this tension implies a value to be embraced, even if the experience involves discomfort, uncertainty and shifts of attention. This tension between expectations and reality indicates an awareness of mental activity, which might be perceived as uncomfortable noise. If self-exploration is characterised by uncertainty, but it also includes the possibility for a choice, then silence-phenomena are highly related to *eudemonia*. This is to say that eudemonic happiness manifests itself as the virtue of deciding wisely and making room for freedom instead of merely responding to lifés constraints. This aspect of *eudemonia* corresponds as well to the inner human life in terms of dreams, feelings and thinking.

Inner silence, Eudemonia and virtues

When it comes to *eudemonia*, it is closely related to the Greek term *areté*, which can be translated as "virtue", which predominantly refers to a high moral standard. In line with this, values seem to be at the core of virtue. The most important trait of a value that guides virtues is that it is supposed to be good. On this basis, "virtue" may very well be defined in functional terms: "A *good* thing of a certain kind is that which has the *virtues* that enable it to carry out its *function* well" (Pakaluk, 2005, p. 6). This brings us directly to the core of how Aristotle defined the term *eudemonia* – by referring to the ultimate goal of a human life, namely, to being in accordance with what one is meant to be. In this sense, *eudemonia* is understood in a teleological perspective; that the fulfilment of one's life's goal is the ultimate understanding of happiness. *Eudemonia* also had some divine connotations as this teleological perspective on the meaning of life reflects a purpose and meaning that lies behind an individual's life per se. Therefore, already in ancient times, the term was associated with blessedness and "being blessed by a spirit or god" (Pakaluk, 2005, p. 47). Another Greek term for being blessed is *makarios*.

However, "*eudaimonia* seems to be mental tranquility alone while that of *makarios*, mental tranquility plus freedom from physical pain" (Miller, 2015, p. 174).

The relationship between the mental and physical has always been a problem for psychology. Wilhelm Wundt solved this with the principle of parallelism, whereby "[a]n interrelation between sensations and *physiological* stimuli must necessarily exist, however, in the sense that different kinds of stimulation always correspond to different sensations" (Wundt, 1907, p. 49). Physiology is to be regarded as a constraining factor, but there is no demonstrable causality between physiology and mental processes. Vygotsky also shared this perspective influenced by Spinoza (Vygotsky, 1997). Spinoza, in turn, can very well be regarded in the light of the stoics (Miller, 2015). According to the Stoics, *eudemonia* means not only to live in accordance with what one is supposed to be but also in accordance with *physis*, the whole universe (Adkins, 1970). Thus, choices do not have to be completely free, but there has to be the opportunity to make them anyway. Inner silence, therefore, is an opportunity for human beings to explore the contents of consciousness but also to become aware of the relationship between them and different physiological activations. This, following Vygotstky's and Spinoza's premises, can open up a space of freedom from environmental and mental constraints. According to Aristotle, the experience of freedom allows eudemonic happiness to flourish and it is key to our ability to fully embracing such happiness. We will now explore the implications of these theoretical aspects for counselling practices. More particularly, we will examine the experience of inner-silence, be it associated with introspection, prayer, meditation, or contemplation.

Discussion: implications for counselling practice

In the Dictionary of Counselling, silences are defined as the temporary absence of any overt verbal or paraverbal communication between counsellor and client (Feltham & Windy, 2004). In the interaction between client and counsellor silences can be experienced as challenging or disturbing. For example, participants of a counselling group have reported perceiving silences as symptoms of personality traits, such as a lack of social skills, a fear of being evaluated negatively, or as not being able to speak unless asked, and in a few cases as reflecting excitement (Yildirim, 2012). Thus, counsellors need the skills to interpret the silences that appear in individual or group sessions. They also need to be able to promote silent settings as they can become an opportunity for the client to listen to themselves (Mearns, 2003). Counsellors can allow silences to take place in therapeutic settings since this can foster the perception of rapport (Sharpley, Munro, & Elly, 2005). Thus, they can use silences to express empathy, to enhance and facilitate reflection, to challenge the client to take responsibility, and to facilitate the expression of feelings (Ladany, Hill, Thompson, & O'Brien, 2004). However, since silence-phenomena are rooms for a wide spectrum of affective processes (Lehmann, 2018), it can also be the case that counselling trainees experience silences as sources of anxiety (Sharpley, 1997). This can be due to the fact that silence-phenomena act as contrasts in the flow of experience, which can increase uncertainty about the contents of experience or about the future (Lehmann, 2018). In this vein, the impact of silence-phenomena in psychotherapeutic processes needs to be further studied in terms of their influence on affective processes, awareness, and creating room for genuine dialogues (Lehmann, 2014). In this article, we suggest three specific paths for exploring inner silence and for promoting *eudemonia* during secular or pastoral counselling practices.

Inner silence as a source of meditation and contemplation

According to Ricard (2013), a Buddhist monk also known as "the happiest man on earth", the I lives in the present: "It's a locus of consciousness, thoughts, judgment and will" (p. 351). Yet, continues the author, identification with, and attaching to, ideas about the self as permanent and rigid, obstructs our relationship with reality and with the present moment, thereby creating suffering and blocking our experience of authentic happiness. In contrast, participants of mindfulness courses such as MBSR (Mindfulness-Based Stress Reduction) identify with discourses that promote an understanding of the self which is dialogical, flexible, and fluid, and they see meditation as a source of inner dialogues (Mamberg & Bassarear, 2015). Thus, providing counselling trainees with mindfulness courses promotes their experiences of psychical and psychological wellbeing (Christopher & Maris, 2010). At the same time, evidence suggests that when clinicians consistently practice mindfulness themselves it can be beneficial for the psychotherapeutic process and for the clients (Shapiro & Carlson, 2017). Yet, further research is needed when it comes to how to effectively integrate a mindfulness practice in psychotherapy and counselling (Davis & Hayes, 2011).

In a similar vein, the training of meta-cognitive processes (e.g. by mindfulness practices) emphasises that training one's attention is a value-driven process, since it is not just about exercising one's attention, but it is doing so with particular intentions and attitudes in mind, ones that promote discernment and self-compassion (Shapiro, Carlson, Astin, & Freedman, 2006). That is, training attention as a tool for observing different I-positions through meta-positions or to engage in various realms of experience through de-positioning requires the further activation of specific affective processes. For example, a further benefit of mindfulness practices as a realm of silence-phenomena is that they promote an awareness of beauty, positive experiences such as joy, as well as moments that facilitate insight or growth during the psychotherapeutic process (Shapiro & Carlson, 2017). Research on silence-phenomena in an educative setting has shown similar effects, as students perceived silent experiences as promoting an awareness of "poetic instants" – moments of beauty and the temporary reconciliation of the tensions of life, experiences that are associated with existential insights (Lehmann, 2018). These existential insights can afford the person in the counselling room to feel human and the ability to cope with uncertainty and tensions through healthier attitudes (Shapiro & Carlson, 2017; Lehmann, 2018).

Inner silence and prayer

Silence is an important part of the spiritual tradition of the East and the West, and it has been from ancient times. For instance, Newby (2011) supports the idea of a spiritual (but not necessarily religious) approach to happiness, wherein the central idea is to develop the psyche so that it might find itself and thereby "win" in the arena of life. However, seen as an explicitly religious approach, this can lead to such questions as: *Can those who do not know inner silence ever attain truth, beauty, or love? Do not wisdom, artistic vision, and devotion spring out of inner silence, where the voice of God is heard in the depths of the human heart?* (Sarah & Diat, 2017). In the context of spiritual or religious counselling, such as Eastern Orthodox traditions, inner silence or stillness are placed at the centre of prayer (St. Nikodimos of the Holy Mountain & St. Makarios of Corinth, 1983). At the same time, prayer is treated in spiritual traditions as a crucial practice to achieve and embrace well-being both for the counsellor and the client (Gubi, 2007).

Thus, inner silence is associated with the concept of consciousness, which at the same time can be related to the notion of the soul. The soul, or the inner self, suggests a bridge between psychological and spiritual dimensions. Consequently, considerations of the notion of the spirit involve the ethical idea of the possibility of tuning one's voice to what is genuine or wise in particular situations, that is, if one learns to listen to the voices within the self. Thus, the idea of inner-silence implies being aware of, perceiving, thinking of, and analysing the contents that show up in consciousness. This suggests a constructive movement from mere interactions among I-positions to a genuine dialogue within the self. Indeed, to speak about a "Silent consciousness" (Lanza & Berman, 2009) may sound paradoxical, yet the connection between inner silence and wisdom is widely asserted in Asian, Western, Middle Eastern and shamanic traditions. That is, even if contemplative practices differ across cultures, "inner silence" is often a goal shared by them (Baars, 2013). This inner silence recalls a movement between the realms of self-exploration and those of existential encounters with others, where relational depth blossoms (Lehmann, 2018).

From a religious perspective, the absence of ambient noise can promote listening to God´s voice. Indeed, the most widespread spirituality in early Christianity (starting in the fourth century) fostered the importance of inner silence. *Hesychasts* (*hesychasm*, meaning "quiet", "silence" and *isi'xazo*, meaning "to be still") believed in two types of consciousness: ego-centred and ego-transcendent. Ego-centred consciousness is a state dominated by attachments to the senses, emotions, intellect, and imagination, whereas ego-transcendent consciousness involves detachment from those faculties (e.g. de-positioning in terms of Dialogical Self Theory). The shift from ego-centred to ego-transcendent consciousness is called *metanoia* in Greek (Liester, 2000). Underlying this idea of the direction of the ego are ethical premises that involve a movement of the self. The notion of an I-position (Hermans, Kempen, & Van Loon, 1992; Hermans, 2001) speaks to the multiple perspectives that the self can have, and recognises the existence of diverse voices of the self. In a similar manner, we suggest considering prayer as a tool that can be used to confront different I-positions by bringing them into dialogue and by tuning them to more clearly hear the voice that feels most genuine from an ethical perspective. These considerations can be understood in psychological terms (e.g. as part of meta-cognition) or spiritual terms (e.g. in terms of ethics, religion, or secular understandings of spirituality). Such a definition speaks to the possibility of embracing happiness by searching for freedom in one's life.

Conclusion

In this article, we have provided a number of theoretical distinctions to explain what inner silence and eudemonic happiness are, and we have explained the human search for *eudemonia* as involving the search for inner silence. In terms of Dialogical Self Theory, we, therefore, suggest that the practice of inner silence can increase awareness of the content of, and the dynamics between, I-positions. This awareness can, in turn, empower the person to feel more freedom over their thoughts, feelings and actions. In addition to this, we suggest that meditation, contemplation and prayer, can be used in the context of counselling ("prayer" for those who follow a particular religion, "meditation" as associated with secular mindfulness practices and religious meditations, and "contemplation" as part of spiritual practices, or daily moments of appreciation). These practices can help to promote the awareness and internalisation of the dialogical nature of the self, and the experience of genuine dialogues with oneself or with others. Both for the counsellor and the client, these silent experiences can serve as phenomenological spaces that enable the person to immerse in, or to detach attention from, a particular focus. This can alter the reactions of the person and can create room for freedom within the constraints of the given circumstances.

More research that focuses explicitly on integrating Dialogical Self Theory in such contexts could help to better understand the dynamics of the different positionings of the self before, during, or after embracing inner silence. Such research could more clearly demarcate the tensions between different I-positions and could allow us to better understand how to promote genuine dialogues. Entering the room of silence-phenomena can open up the possibility of detaching from, and otherwise transcending, different I-positions. This creates the possibility of exercising freedom from the constraints of experience, and this freedom that produces the experience of meaningful completeness can create an experience of eudemonic happiness. Such genuine happiness paves a path for authenticity. In other words, it involves the process of fine-tuning one's inner voice and embracing dialogues with oneself, with others and (or) with God.

Acknowledgements

We thank Lucas Mazur for proofreading the article and for his valuable comments on how to improve our arguments.

Disclosure statement

No potential conflict of interest was reported by the authors.

References

Adkins, A. W. H. (1970). *From the many to the one. A study on personality and views on human nature in the context of ancient Greek society, values and beliefs.* London: Constable.

Baars, B. (2013). A scientific approach to silent consciousness. *Frontiers in Psychology, 4,* 678. doi:10.3389/fpsyg.2013.00678

Bruneau, T. J., & Ishii, S. (1988). Communicative silences: East and West. *Word Communication, 17*(1), 1–33.

Buber, M. (1950). *I and thou.* Edinburgh: T & T.

Cage, J. (1939/1961). *Silence. Lectures and writings by John Cage.* Middletown: Wesleyan University Press.

Christopher, J. C., & Maris, J. A. (2010). Integrating mindfulness as self-care into counselling and psychotherapy training. *Counselling and Psychotherapy Research, 10*(2), 114–125. doi:10.1080/14733141003750285

Colum, K. (2011). *The power of silence: Silent communication in daily life.* London: Karnac Books.

Cooper, M. (2004). Encountering self-otherness: I-I and I-me modes of self-relating. In H. Hermans, & G. Dimaggio (Eds.), *The dialogical self in psychotherapy* (pp. 60–73). Hove: Brunner-Routledge.

Cooper, M. (2012). Experiencing relational depth: Self-development exercises and reflections. In R. Knox, D. Murphy, S. Wiggins, & M. Cooper (Eds.), *Relational depth: New perspectives and developments* (pp. 137–152). Basingstoke: Palgrave Macmillan.

David, S. (2016). *Emotional agility: Get unstuck, embrace change and thrive in work and life.* New York, NY: Penguin.

Davis, D. M., & Hayes, J. A. (2011). What are the benefits of mindfulness? A practice review of psychotherapy-related research. *Psychotherapy, 48,* 198–208.

Eriksen, T. B. (1972). *Den greske filosofi.* Oslo: Gyldendal.

Feltham, C., & Windy, D. (2004). *Dictionary of counselling* (2nd ed.). London: John Wiley & Sons.

Frankl, V. (1966/2000). *Fundamentos y Aplicaciones de la Logoterapia.* Buenos Aires: San Pablo.

Gubi, P. M. (2007). *Prayer in counselling and psychotherapy: Exploring a hidden meaningful dimension.* London: Jessica Kingsley Publishers.

Hermans, H. J. M. (2001). The dialogical self: Toward a theory of personal and cultural positioning. *Culture & Psychology, 7*(3), 243–281.

Hermans, H. J. M. (2003). The dialogical self: Toward a theory of personal and cultural positioning. *Journal of Constructivistic Psychology, 6*(2), 89–130.

Hermans, H. J. M., & Hermans-Konopka, A. (2010a). *Dialogical self theory: Positioning and counter-positioning in a globalizing society.* Cambridge: Cambridge University Press.

Hermans, H., & Hermans-Konopka, A. (2010b). Positioning theory and dialogue. In H. Hermans, & A. Hermans-Konopka (Eds.), *Dialogical self theory. Positioning and counter-positioning in a globalizing society* (pp. 120–199). Cambridge: Cambridge University Press.

Hermans, H. J. M., Kempen, H. J., & Van Loon, R. J. (1992). The dialogical self: Beyond individualism and rationalism. *American Psychologist, 47*(1), 23–33.

Kim, M., Chang, S. I., Seong, J. C., Holt, J. B., Park, T. H., Ko, J. H., & Croft, J. B. (2012). Road traffic noise: Annoyance, sleep disturbance, and public health implications. *American Journal of Preventive Medicine, 43*(4), 353–360.

Kurzon, D. (1998). *Discourse of silence.* Amsterdam: John Benjamins.

Ladany, N., Hill, C. E., Thompson, B. J., & O'Brien, K. M. (2004). Therapist perspectives on using silence in therapy: A qualitative study. *Counselling and Psychotherapy Research, 4*(1), 80–89.

Lanza, R., & Berman, B. (2009). *Biocentrism: How life and consciousness are the keys to understanding the true nature of the universe.* Dallas, TX: BenBella Books.

Lazzari, C. (2009). *Spiritual counselling in medicine: Theories and techniques of counselling during stressful life events, severe illnesses, and palliative care.* New York, NY: iUniverse.

Lehmann, O. V. (2014). Towards dialogues with and within silence in psychotherapy processes: Why the person of the therapist and the client matters? *Culture & Psychology, 20*(4), 537–546. doi:10.1177/1354067X14551298

Lehmann, O. V. (2016). Something blossoms in between silence-phenomena as a bordering notions in psychology. *Integrative Psychological and Behavioral Science, 50*(1), 1–13. doi:10.1007/s12124-015-9321-7

Lehmann, O. V. (2018). *The cultural psychology of silence. Treasuring the poetics of affect at the core of human existence* (Unpublished doctoral dissertation). NTNU Norwegian University of Science and Technology, Trondheim, Norway. Retrieved from https://brage.bibsys.no/xmlui/handle/11250/2484902.

Lehmann, O. V., & Klempe, S. H. (2017). The musicality of poetry and poetic musicality: A case of cultural psychology approach to study the creativity within emotions and meaning. In O. V. Lehmann, N. Chaudhary, A. C. Bastos, & E. Abbey (Eds.), *Poetry and Imagined Worlds: Creativity and everyday experience* (pp. 175–195). London: Palgrave Macmillan.

Lehmann, O. V., & Valsiner, J. (2017). Deep feelings in actions: Where cultural psychology matters. In O. V. Lehmann, & J. Valsiner (Eds.), *Deep experiencing. Dialogues within the self* (pp. 93–102). New York, NY: Springer.

Liester, M. B. (2000). Hesychasm: A Christian path of transcendence. *Quest, 65,* 54–59.

Mamberg, M. H., & Bassarear, T. (2015). From reified self to being mindful: A dialogical analysis of the MBSR voice. *International Journal of Dialogical Science, 9*(1), 11–37.

Mearns, D. (2003). *Developing person-centred counselling.* London: Sage.

Mearns, D., & Cooper, M. (2005). *Working at relational depth in counselling and psychotherapy.* London: Sage.

Miller, J. (2015). *Spinoza and the stoics.* Cambridge: Cambridge University Press.

Newby, M. J. (2011). *Eudaimonia – happiness is not enough.* Leicester: Troubador.

Nir, D. (2016). Becoming the leader of your decisions. In H. Hermans (Ed.), *Assessing and stimulating a dialogical self in groups, teams, cultures and organizations* (pp. 1–18). Switzerland: Springer.

Orlandi, E. P. (1995). *As formas do silêncio: No movimento dos sentidos.* Campinas: Unicamp.

St. Nikodimos of the Holy Mountain and St. Makarios of Corinth. (1983). *The Philokalia: The complete text* (Vol. 1). In G. E. H., Palmer, P. Sherrard, & W. Kallistos (Eds.). London: Faber & Faber.

Pakaluk, M. (2005). *Aristotle's Nicomachean ethics: An introduction.* Cambridge: Cambridge University Press.

Puchalska-Wasyl, M. (2010). Dialogue, monologue, and change of perspective – three forms of dialogicality. *International Journal for Dialogical Science, 4*(1), 67–67.

Ricard, M. (2013). A Buddhist view of happiness. In I. Boniwell, S. David, & A. Conley Ayers (Eds.), *The Oxford handbook of happiness* (pp. 344–356). Oxford: Oxford University Press.

Sarah, R., & Diat, N. (2017). *The power of silence: Against the dictatorship of noise.* San Francisco, CA: Ignatius Press.

Schmid, P. F. (2013). Dialogue as the foundation of person-centered therapy. In R. Knox, D. Murphy, S. Wiggins, & M. Cooper (Eds.), *Relational depth. New perspectives and developments* (pp. 155–174). London: Palgrave Macmillan.

Shapiro, S. L., & Carlson, L. E. (2017). Mindfulness and self-care for the clinician. In S. L. Shapiro, & L. E. Carlson (Eds.), *The art and science of mindfulness: Integrating mindfulness into psychology and the helping professions* (pp. 115–126). Washington, DC: American Psychological Association.

Shapiro, S. L., Carlson, L. E., Astin, J. A., & Freedman, B. (2006). Mechanisms of mindfulness. *Journal of Clinical Psychology, 62*(3), 373–386.

Sharpley, C. F. (1997). The influence of silence upon client-perceived rapport. *Counselling Psychology Quarterly, 10*(3), 237–246.

Sharpley, C. F., Munro, D. M., & Elly, M. J. (2005). Silence and rapport during initial interviews. *Counselling Psychology Quarterly, 18*(2), 149–159.

Tangene, C. (2017). Feeling the silence of the ocean: Experiencing the night shift. In O. V. Lehmann, & J. Valsiner (Eds.), *Deep experiencing. Dialogues within the self* (pp. 39–50). New York, NY: Springer.

van Loon, R., & van den Berg, T. (2016). Dialogical leadership. The "Other" way to coach leaders. In H. Hermans (Ed.), *Assessing and stimulating a dialogical self in groups, teams, cultures and organizations* (pp. 75–94). Switzerland: Springer.

Vygotsky, L. S. (1997). *Problems of the theory and history of psychology. The collected works of L. S. Vygotsky.* In R. W. Rieber, & J. Wollock (Eds.). New York, NY: Plenum Press.

WHO. (2017). *Depression: Let's talk" says WHO, as depression tops list of causes of ill health.* Retrieved from http://www.who. int/news-room/detail/30-03-2017-depression-let-s-talk-says-who-as-depression-tops-list-of-causes-of-ill-health

Wong, P. (2011). Positive psychology 2.0: Towards a balanced interactive model of the good life. *Canadian Psychology*, *52*, 69–81.

Wundt, W. (1907). Outlines of psychology (3rd rev. English ed. from 7th rev. German ed.) (C. H. Judd, Trans.). Leipzig, Germany: Wilhelm Engelmann. http://dx.doi.org/10.1037/12406-000.

Yildirim, T. (2012). The unheard voice in group counselling: QUIETNESS. *Educational Sciences: Theory and Practice*, *12*(1), 129–134.

Creating space for happiness to emerge: the processes of emotional change in the dialogical stage model

Georgia Gkantona

ABSTRACT

It is very often in counselling and psychotherapeutic procedures that clients need to cope with negative emotions such as sadness, guilt, fear or anger. The implementation of the Dialogical Stage Model (DSM) (Hermans, H. & Hermans-Konopka, A. [2010]. *Dialogical self theory: Positioning and counter-positioning in a globalizing society*. New York: Cambridge University Press) can sustain these kinds of emotional processes. It comprises seven stages that entail the articulation, clarification and change of emotions according to a dialogical procedure. The therapeutic procedure is conceived as a meeting of living persons engaged in a collaborative dialogue. These dialogical processes motivate the creative resources of the client and delineate new emotional relationships, aiming at the creation of space in the self for positive emotions of happiness, hope, affection, intimacy and love to emerge. The therapist encounters the challenge to be fully present in the moment, as a comprehensive embodied living person (Seikkula, J. [2011]. Becoming dialogical: Psychotherapy or a way of life? *Australian and New Zealand Journal of Family Therapy*, 32(3), 179–270). Thus, focusing on his/her experiencing during the session (Rober, P. [2008]. The therapist's inner conversation in family therapy practice: Struggling with the complexities of therapeutic encounters with families. *Person-Centered and Experiential Psychotherapies*, 7[4], 245–278) contributes to acquiring better understanding of what goes on in the clients' lives. Moreover, sharing this experience with the client contributes to an emotional extending of the client's self. A clinical case example is used as an illustration of these kinds of therapeutic dialogical processes.

Introduction

The need to elaborate on negative emotions is a very frequent task in clinical practice. Regarding Anxiety Disorders, fear and distress have been specified as two negative affect traits associated to different degrees with all related categories (Carvalho et al., 2014). Thus, coping with these emotions is often of major importance in order to accomplish positive outcomes. Empirical data show that therapeutic procedures focusing on emotional processes seem to be effective for anxiety and worry treatment in general (Borkovec, Alcaine, & Behar, 2004; Campbell-Sills & Barlow, 2007; Mennin, Heimberg, Turk, & Fresco, 2002).

Emotion processes presuppose discriminating emotions between both primary and secondary emotions, and between emotional experience that is adaptive or maladaptive (Fosha, 2000). Primary emotions are the person's most fundamental direct initial reactions to a situation. Such examples can be feeling happy due to a success or experiencing intense fear due to a sudden pain in the chest. Secondary emotions are responses to one's thoughts or feelings rather than to the situation, such as being afraid something bad will happen that will destroy my happiness or feeling anxious about experiencing intense fear (Greenberg, 2010, 2011; Greenberg & Pascual-Leone, 2006).

Primary emotions may be adaptive and thus can be accessed for their useful information as well as they may be maladaptive and need to be transformed. Maladaptive emotions are familiar feelings that occur repeatedly, such as the anxiety that someone experiences when being in an insecure relationship, feelings of wretched worthlessness, or shameful inadequacy that plague one's life. These maladaptive feelings neither change in response to changing circumstances nor provide adaptive directions for solving problems when they are experienced. Primary adaptive emotions need to be accessed in therapy as they contain adaptive information and help the client to organise his/her further action, whereas maladaptive emotions need to be accessed and regulated in order to be transformed. Secondary emotions need to be bypassed to get to more primary emotions (Greenberg, 2010, 2011; Greenberg & Pascual-Leone, 2006). Moreover, a discrimination between emotion and feeling can be that emotion is a broader concept that includes neurophysiological aspect, whereas feeling refers to the subjective experience of the emotion (Hermans & Hermans-Konopka, 2010).

An emotion process that may assist positive outcomes in counselling or therapeutic settings is changing an emotion with another emotion (Greenberg, 2008). This kind of change in emotional experience is stimulated in the beginning by activating the maladaptive experience of the negative emotion. This happens when the therapist invites the client to enter in the negative emotion that suffers from and talk about the way this emotion influences his/her life, relations, physical functioning etc. Then the client is helped to access other negative emotions that are related to this maladaptive experience of this negative emotion. What follows is accessing a client's adaptive emotions, and vitalising this way a more resilient sense of self, so as to help transform the person's maladaptive emotions and related beliefs (Greenberg, 2004, 2012). In this way the new self-experience and views are integrated with the existing negative experience and a new self-organisation is succeeded. Thus, a process of accessing the adaptive and bringing it into contact with the maladaptive helps transform the maladaptive schemes (Greenberg, 2012; Greenberg & Pascual-Leone, 2006).

This process does not imply that negative feelings are replaced with happy feelings just by trying to look on the bright side, but by the evocation of a meaningfully embodied alternate experience so as to help the client to move from the maladaptive experience caused by the negative feeling. For emotional change to be facilitated, clients also need to develop new narratives which assimilate that new experience into existing cognitive structures and generate new ones. Therapy thus involves changing both emotional experience and the narratives in which they are embedded (Greenberg, 2012; Greenberg & Angus, 2004). An integration of the above mentioned processes is described in the Dialogical Stage Model that follows.

A dialogical stage model (DSM) for changing emotions

In the context of the Dialogical Self Theory (Hermans & Hermans-Konopka, 2010) emotions are considered to be embedded in a social context and result from real, imagined, anticipated or recollected outcomes of social relationships. Elaborating on this view, it is argued that emotions function as part of an organised position repertoire and essentially contribute to the processes of dialogue either in the self of an individual or between individuals. Thus, understanding emotions can be helped by considering them in bi-directional relationships to the self: emotions influence and organising the self as well as the self can confirm, allow and change emotions. Moreover, it is proposed to understand emotions as movements in a metaphorical space in which one places oneself or is placed in relation to somebody or something.

On this theoretical basis, Hermans and Hermans-Konopka (2010) outlined a seven-stage model for changing emotions conceived of as dialogical movements in a metaphorical space. According to this model that is described here, as presented in the book "*Dialogical Self Theory: Positioning and Counter-positioning in a globalizing society*," the person can go into an emotion, leave the emotion, go to another emotion that is a counter-emotion, leave this counter-emotion and develop dialogical relationships between them. These dialogical movements between emotions stimulate processes that create space for new emotions to emerge, promote the self to higher levels of development and enhance the individual's functioning.

Stage 1: Identifying and entering an emotion

In order to identify an emotion that is important in a certain period of someone's life, emotional awareness is a prerequisite. Emotional awareness is defined as the person's ability to recognise and describe his or her own emotions especially when they are mixed and complex (Lane & Schwartz, 1987). It helps the person to make sense of emotional experience and to overcome avoidance.

Awareness involves approaching and accepting emotions. Acceptance of emotional experience, as opposed to its avoidance, is the first step in awareness work. Having accepted the emotion rather than avoiding it, the therapist then helps the client in the utilisation of emotion. Clients are helped to make sense of what their emotion is telling them and to identify the goal, need, or concern that it is directing them to attain. Emotion is thus used both to inform and to move (Greenberg, 2006, 2011).

In therapeutic and counselling procedures, the initially identified emotions that clients want to change are usually negative ones. As clients suffer from them they want to get rid of them quickly. However, in this attempt, they often acknowledge either the potentially adaptive sides of these negative emotions or the important messages for changing client's self or his/her relationships towards more functional ways.

Entering an emotion is essential in order to get acquainted with its experiential quality and understand it "from the inside". When people respond to emotions in rational ways, the awareness and accessibility of the emotion is often blocked and the dialogical movements are restricted. Entering the emotion allows the person to become immersed in it and to speak through the emotion rather than about it.

Stage 2: Leaving the emotion

In order for dialogical movements to proceed, a person must be able to leave the emotion that he/she was previously immersed in. This dis-identification process depends on the level of flexibility one has to move from one position to another (Hermans & Hermans-Janssen, 1995). According to the authors of the model, the way in which positions are organised in the repertoire of the self (e.g. in opposition or coalitions) is what influences the way that a person moves more or less flexibly from one position to another.

Stage 3: Identifying and entering a counter-emotion

Another emotion that is clearly different from the initial one and can be seen by the participant as a "helpful response" to the initial emotion is identified. The introduction of this counter-emotion is conceived to be able to change an emotion that the client needs to change. Particularly, in cases of initial maladaptive emotions such as sadness, the introduction of a counter-emotion (e.g. hope) can be a promising starting point for influencing and transforming the initial one.

Stage 4: Leaving the counter-emotion

A similar process to stage 2 is introduced by inducing the client to leave the counter-emotion. Usually the counter-emotion may be valuable but is not as dominant as the initial one and thus clients leave it more easily. This stage also has a crucial transient functionality as it establishes important dialogical movements.

Stage 5: Developing dialogical relations between emotion and counter-emotion

Emotional change can take place when the person succeeds in synthesising opposite emotions, described as the "dialectical synthesis of opposite schemas" (Greenberg, 2004). Thus, coping with a negative emotion can be accomplished by the introduction of new elements to an activated emotional structure. Therefore, in this stage model it is proposed that an emotion can be changed when it receives an innovative dialogical impulse from a counter-emotion.

A cornerstone of the dialogical self theory is that the self consists of a multiplicity of positions. Nevertheless, when a person is stuck in a dominant emotional position (e.g. sadness) and cannot move flexibly to others, dialogical movements are not processed within the self. Then, the sense of self is fused within the experience of this negative emotion and the person feels imprisoned in sadness. Thus, dialogical movements between emotions take place only if the self can dis-identify from a space absorbing position. In order for the self to proceed, dis-identification experiencing space beyond a dominant emotion is necessary.

Stage 6: Creating a composition of emotions

The therapist induces the client to create a composition of emotions. The notion of composition reflects the conceptualisation of the self as an organised position repertoire and the essential quality of such a composition is its pattern. Emotions take their place in a larger whole and the client is induced to reflect on this according to formed constellations, coalitions, central or peripheral positioning etc. as well as the extracted meaning of such placements.

The composition of emotions as a pattern, where a great number of positions are brought together, enables the person to take a meta-position. A meta-view is allowed as the client reflects his or her patterned emotional positions from a distance.

In practice, this stage can be implemented using stone patterns different in size, colour, and texture. Stones can motivate the creative resources of the client, as they symbolise the multiplicity of emotional positions in the landscape of the mind, as well as give the client the possibility to process actual dialogical movements among them. This process was further elaborated in composition work (Konopka & Van Beers, 2014).

Stage 7: Introducing promoter positions in the context of emotions

In this final stage, the notion of promoter position is introduced. Two main possibilities are in the application of the model. Firstly, emotions are changed by a particular promoter position that plays a significant role in the person's life. Secondly, particular emotions function as promoter positions themselves. Assuming that some emotions can function as promoters, the following principles apply to their functionality: (a) they imply considerable openness towards the future, (b) through their openness and broad bandwidth, they integrate a variety of new and existing emotions in the self, (c) with their central place in the position repertoire, they have the potential to reorganise the self towards a higher level of development, (d) they function as guards of the continuity of the self and at the same time they leave space for discontinuity.

In this theoretic context, this paper focuses on the description of a clinical case example as an illustration of DSM. Presenting the particular dialogical processes between the client's emotions as well as the outcomes, the aim is to highlight the potential utility of this model in terms of creating space in the self for positive emotions such as happiness to emerge.

Method

The therapist and the therapeutic procedure

This particular therapeutic procedure is conceived as a meeting of living persons engaged in a collaborative dialogue. The therapist follows the lead of the client and actively seeks the nuances of his description. She encounters the challenge to be fully present in the moment, as a comprehensive embodied living person (Seikkula, 2011). Furthermore, the therapist, focusing on her experiencing during the session aims at acquiring better understanding of what goes on in the client's life (Rober, 2008) as well as putting emotions into words such as anger or sadness that the client seems unable to express. She also intends to utilise the therapeutic relationship to generate new emotions: a new, more adaptive emotion will be evoked in response to interactions with the therapist characterised by empathic attunement to affect while providing safety, comfort, and a sense of security (Robinson, Dolhanty, & Greenberg, 2015).

The therapist is the writer of the present article. At the time of the sessions she held a PhD in counselling psychology and a private practice for 11 years. Her basic theoretical orientation had been systemic but over the past years there was a turn to the dialogical perspective in individual and family therapy.

The therapeutic procedure took place in 2015. The illustration of the DSM is based on excerpts from the 3rd to the 7th session, while the whole therapeutic procedure lasted for 15 sessions. Apart from the application of the DSM, the therapeutic procedure was ruled by individual systemic therapy methodology (Bertrando, 1996). It is important to mention that the client's name has been changed to preserve his anonymity as well as other potentially revealing details.

Case illustration

Client description & presenting problem

Jim is a 45-year-old man who studied Business Administration and is working as a civil servant. He lives in his hometown, he is married and he has two school-age children. He got married ten years ago and he says that after his wife gave birth to his second child, family needs increased so much so that there was no time for his own personal needs to be met. As years went by, he was more and more preoccupied with keeping up with everyday duties, and subsequently started feeling anxious and guilty that he was not sufficient in his roles as a professional, husband and parent. In order to cope with his anxiety and the sense that he could not enjoy his life, he started taking psychiatric medication. A year later, he decided that he no longer wanted to be dependent on medication. At present, he has decided that he wants to be happy again and try to cope with his negative emotions by seeking psychotherapeutic treatment.

The implementation of the dialogical stage model

Identifying and entering an emotion

In order to help the client to make progress with his emotional awareness, the therapist wanted to know which was the most difficult or unbearable emotion for him at the time. Jim started talking about his feelings of anxiety that were constantly present and the fact that he could not relax. Reflecting on that anxiety, he mentioned that it was connected with guilt. He felt guilty in different ways. If he was to live his life making himself a priority, he felt guilty about the other family members and considered himself a selfish husband or a bad father. On the other hand, if he was to make fulfilling the needs of others a priority, he felt like an "idiot" that was being exploited by others. Then he would reach a conclusion that he could not find a balance between satisfying personal needs and the needs of others. Consequently, he would end up feeling frustrated and sad about this incapacity. Over the past few years, his life has been tainted by this vicious emotional circle.

He concluded the initial session saying that as his life was being consumed by those kinds of negative emotions, at the end of the day he would feel empty as though everything had drained his energy. As focusing on the body awareness promotes emotional awareness (Sze, Gyurak, Yuan, & Levenson, 2010), the therapist invited him to enter a process of further describing and recognising the characteristics of this maladaptive emotion by posing exploratory questions about the embodied experience accompanying this emotion. He identified

> This feeling of emptiness is like a hole in my chest … and it makes my body heavy and rigid. It causes a sense of distress that extends to every part of my body. Then I feel like being trapped in a loop … and the funny thing you know is that I lead myself into this situation. But I don't know how to stop it or escape from it (he is puffing) … . This sense of emptiness is really awful!

Then the therapist wondered in a reflective way about the message his body was sending to him while experiencing this sense of emptiness. Jim made a reference to an emotional experience of his childhood providing an opening to his relationships among his family members of origin: He had felt that he had to be the "good boy" so as not to add "trouble" to his family. It was a period during which his family was facing the antisocial behaviour of his older brother. There were often quarrels between his parents and his brother. In order to protect himself from that negative emotional atmosphere he tried not to feel anything. Moreover, in order to avoid creating additional concerns to his parents, he restricted his emotional needs and didn't ask for anything, trying to be invisible. But nowadays his body screams: "I can't stand being invisible anymore!"

Leaving the emotion

Then the therapist invited Jim to leave the emotion of emptiness and imagine his life without it. He replied that the way he experiences everyday life is coloured by this emptiness to the extent that he feels that it is not possible to differentiate one from another any more. Then the therapist further assisted Jim to leave this emotion using a variation of the miracle question (de Shazer et al., 2007; Stith et al., 2012) so as to lift him out of his constraining saturated view in which everyday life is experienced in emptiness. The aim was to open space for the development of a possible vision of what his life would be like in the absence of this maladaptive emotion:

> Suppose tonight while you are sleeping, a miracle happens and this emotion of emptiness leaves. You are asleep so, of course, you don't know that the miracle has happened. When you wake up the next morning, what would be the first thing that would tell you a miracle had occurred?

He said that a smile could be the first thing that would tell him a miracle had occurred. Afterwards he started describing a possible vision of life without the feeling of emptiness. In this way he moved to a perspective of detaching from the emotion of emptiness.

Identifying and entering a counter-emotion

While Jim was talking about a possible perspective of his life without feeling empty, the therapist focused on her own internal experience of that moment. The emotionally positive atmosphere of that moment created feelings of hope and optimism within the therapist. By sharing this internal experience with Jim, the therapist encouraged him to move towards a positive emotional position. Moreover, their therapeutic dialogue was enriched in an emotional way.

At that point, she invited him to recall a totally different emotion from the emotion of emptiness he had experienced in his life. After remaining silent for a while, he narrated a story from college life. One day, he and his best friend wanted to take a trip but couldn't decide where to go. So they thought about going to the train station and taking one of the next trains that had an appealing destination. So that is what they did and he had one of the best holidays in his life. He explained that what he did in that case was to trust himself to reach a decision together with someone else, feeling good at the same moment. And this related to feelings of joy and serenity.

Regarding the therapeutic course of the case upon the completion of this stage, a useful feature that emerged was that Jim went through an emotional process delineating a dialogical movement from a negative and maladaptive emotion to a positive and more adaptive one. This resulted in the establishment of an experiential space where future dialogical processes among different parts of his self could be hosted.

Leaving the counter-emotion

Focusing on this experiential space, the therapist invited Jim to reflect on the process of moving from experiencing a negative emotion to a positive one. In this way, he left the emotional position of serene and joy, focusing on this experiential space where multiple emotional I- positions can be hosted and related to each other.

Developing dialogical relations between emotion and counter-emotion

After acknowledging the possibility of multiple emotions in his emotional terrain, Jim was invited to enter his sense of emptiness again and further contemplate on his perspective of life from this position. He was also asked to identify an image or symbol representing this experience. He said that staring at his life from the bastion of feeling empty is like taking part in a war with no meaning for him. It's like a soldier who has no power or aim in his life. When asked what the voice of emptiness had to tell him, he shook his head in a gesture of self awareness replying "It says you have lost yourself, you cannot hear yourself!"

Then he was invited to enter his emotion of serene joy and represent it in a related image. He came back to that image of being on a train and going to a desired destination. He smiled and his body became more relaxed. When asked what serene joy had to tell him, he said "Stop being afraid of others, don't beat yourself up anymore. Love yourself and listen to yourself more than you do!" The therapist asked Jim what he would answer to that voice and he replied "Enough with this. You have to change!"

Afterwards, the therapist reflected on her own embodied experience stimulated by the ongoing emotional processes which helped Jim to get in touch with an underlying feeling of frustration and anger. These feelings proved to be difficult for Jim to relate to. The important message of his anger to himself further processed the dialogical movements and resulted in the emersion of more emotions such as the longing to change.

Creating a composition of emotions

At this stage the therapist provided Jim with a basket of stones and paper stickers. She asked him to write on different stickers the names of the important positive and negative emotional positions which emerged in the previous stages of the sessions. After identifying a stone for each emotion, he stuck the emotions written on the stickers on the stones. He selected: "I as anxious", "I as guilty", "I as sad", "I as empty", "I as joyful", "I as longing to change" and finally, "I as serene". Thereafter, he had to make a stone constellation that expressed the way he felt at the time. In the centre of his composition he positioned the "I as sad" and next to it the "I as empty". The positive emotions were more peripheral. Dialogical movements among the emotions were promoted by questions such as: "Which of the positive emotions has the power to move sadness?" "How can this movement reorganise the other emotions?" The "I as longing to change" proved to be the most powerful emotion that resulted in a pattern with the positive emotions moving closer to the centre.

A collaborative reflection between the therapist and the client facilitated a series of dialogical movements among these emotional positions that resulted in the entrance of a new important embodied experience: "I as internally more flexible". The movements and successive constellations led Jim to that innovative internal experience that can be described as "It's like a ball made of clay inside me that suddenly broke. Now I feel more at ease, more flexible!" A new emerging constellation after the entering of this new emotion inspired him to comment: "You know, this is like exercising control over my emotions!" and he laughed. Discussing it, he admitted that gaining self - mastery would help him to lead his life in a more functional way.

Introducing promoter positions in the context of emotions

After all the processes of emotional change, Jim concluded that his emotional position "Me as internally more flexible" had great potential for creating an internal space for happiness to emerge in his life. Reflecting on this innovative internal experience, he mentioned that being flexible enables him to gain balance among hearing the many voices of his emotional needs as well as important needs of others. For large periods of time, he used to hear only one predominant voice (e.g. what my family wants me to do) while at the same time he felt guilty for not listening to the voice of his personal needs. That guilt was slowly pushing him to the opposite pole of hearing only the voice of his personal needs and expressing anger to his family for the oppression he had gone through. This led to a mood circle in which he could not feel happy in. Thus, positioning himself in the "I as internally more flexible" promotes an internal polyphony that welcomes the dialogue between the needs of many voices simultaneously. He added that this internal flexibility makes him more serene as he feels more capable of regulating his opposing emotions and thus functioning in a better way.

Therapeutic outcomes

With the completion of the DSM processes, short term positive outcomes were obvious during the 7th and the 8th session. Jim mentioned that he had some moments of fun with his family. He was able to relax with them for the first time after many years. It is important to him that they prepared a family dinner all together and he acted in a way that deterred quarrels. He also started talking with his wife about the things they had to do in the home. He felt really happy about these improvements in his relational functioning.

He added that he has been experiencing something new in his everyday life:

> Up to now, when I faced a difficulty in my everyday life, I often felt like being totally blocked and I could not function in a way to cope with the problem. Now I act faster. There were situations that previously paralyzed me, while now I see that I can think about them and have ideas about how to cope with them.

In the following sessions he referred to the experience he had during the emotion composition stage that he described as "a clay ball inside me that broke." He said that this was a turning point in the way he experiences himself. After that release his life now seems easier. So months later after he had reflected on his therapeutic course, he mentioned that putting his inner difficulties with precious experiences of his life into dialogue helped him to be more accepting. That is, he acknowledged both the positive and the negative sides that life has brought on to him and now holds a more comprehensive life perspective. Moreover, these emotional processes helped him to gain a balance among his emotions and be more able to manage them.

Discussion

The presentation of this case example shows the way in which the co-activation of the adaptive feelings of serenity and joy along with or in response to the maladaptive feeling of emptiness resulted in achieving positive outcomes. It seems that transforming one emotion into another counter emotion have the potential of creating space in the self for positive emotions to emerge (Greenberg, 2011, 2012). This transformation was accomplished through the dialectical synthesis of the above opposing schemes. This resulted in compatible elements from the co-activated schemes that formed new higher level schemes. Schemes of different emotional states similarly are synthesised to form new integrations (Greenberg & Pascual-Leone, 1995). Moreover, the dialogical processes of DSM stimulate the integration of emotion and cognition by introducing verbal with non-verbal material together. This particularly happens in the stage of stone patterns.

The dialogical movements among his emotional positions and the successive constellations during the composition stage also motivated Jim's creative resources and helped him to overcome the rigidity that characterised his everyday functioning. The internal flexibility that emerged through this insightful moment characterised as "a clay ball inside me that broke", led him to a more flexible way of living. Possibly, this rigidity was preserved by the monological presence of the "emptiness feeling" that dominated the emotional terrain of his self for a long period. The entering of this polyphonic coexistence introduced by the emotional constellation and their successive movements helped him to move to a more pluralistic sense of the self (Hermans & Hermans-Konopka, 2010) that was experienced as internal flexibility.

During the above mentioned processes, the therapist used her embodied experience so as to better understand Jim's feelings. By reflecting on it, she helped him to get in touch with some feelings such as anger and frustration that were difficult for him to connect with. As part of Jim's extended self, the therapist belongs to the extended domain of Jim's self. Thus, the therapist's reflection as an important element of the therapeutic procedure (Rober, 2008) can be identified as Jim's self-reflection in the extended sense of the term. In this way, the client is helped to extend the external domain of his self and further develop dialogical relation between his internal position and external position (Hermans & Hermans-Konopka, 2010).

Regarding the emergence of the positive emotions in Jim's life and his talking about feeling good after many years adds an important indication related to the potential of the DSM to create space for happiness to emerge, as the subjective well-being is an important facet of happiness (Jacobsen, 2007). Moreover, positive changes in Jim's relational functioning support the usefulness of the applied dialogical processes, as it is argued that happiness is not only feeling good, but doing good also (Di Tella & MacCulloch, 2006). An abundance of empirical literature shows that people who are happier achieve better outcomes in life, including financial success, supportive relationships, strong mental health, effective coping methods, and even physical health and longevity. Moreover, prospective and longitudinal studies show that happiness often precedes and predicts these positive outcomes, rather than simply resulting from them (Lyubomirsky, King, & Diener, 2005). But in what way did the emergence of positive emotions contribute to Jim's functioning better? According to Cohn, Fredrickson, Brown, Mikels, and Conway (2009) positive emotions are the better predictor of whether or not people build important resources and become more satisfied with their lives. Lived experiences like joy and interest are what start the process of exploring, learning, connecting, and ultimately building new resources. Those resources can later improve one's life, offering new opportunities for enjoyment and resource-building.

Finally, it is necessary to say that the presentation of this case example has some limitations. It has a restricted focus on the detailed illustration of the emotional processes of DSM as implemented in a private practice setting. It aims to contribute to the understanding of these processes but is unclear how generalisable this process is in other populations or settings. Although the DSM has a clear structure based on its stages, the processes that emerge with its application can vary due to the particular characteristics and needs of each client.

Disclosure statement

No potential conflict of interest was reported by the author.

References

Bertrando, P. (1996). *Systemic therapy with individuals*. London: Karnac Books.

Borkovec, T. D., Alcaine, O., & Behar, E. (2004). Avoidance theory of worry and generalized anxiety disorder. In R. G. Heimberg, C. L. Turk, & D. S. Mennin (Eds.), *Generalized anxiety disorder: Advances in research and practice* (pp. 77–108). New York: Guilford Press.

Campbell-Sills, L., & Barlow, D. H. (2007). Incorporating emotion regulation into conceptualizations and treatments of anxiety and mood disorders. In J. J. Gross (Ed.), *Handbook of emotion regulation* (pp. 542–559). New York: Guilford Press.

Carvalho, H. W., Andreoli, S. B., Lara, D. R., Patrick, C. J., Quintana, M. I., Bressan, R. A., ... Jorge, M. R. (2014). The joint structure of major depression, anxiety disorders, and trait negative affect. *Revista Brasileira de Psiquiatria, 36*(4), 285–292.

Cohn, M., Fredrickson, B., Brown, S., Mikels, J., & Conway, A. (2009). Happiness unpacked: Positive emotions increase life satisfaction by building resilience. *Emotion, 9*(3), 361–368.

Di Tella, R., & MacCulloch, R. (2006). Some uses of happiness data in economics. *Journal of Economic Perspectives, 20*(1), 25–46.

Fosha, D. (2000). *The transforming power of affect: A model of accelerated change.* New York: Basic Books.

Greenberg, L. S. (2004). Introduction: Emotion special issue. *Clinical Psychology and Psychotherapy, 11,* 1–2.

Greenberg, L. S. (2006). Emotion-focused therapy: A synopsis. *Journal of Contemporary Psychotherapy, 36*(2), 87–93.

Greenberg, L. S. (2008). Emotion and cognition in psychotherapy: The transforming power of affect. *Canadian Psychology/ Psychologie canadienne, 49*(1), 49–59.

Greenberg, L. S. (2010). Emotion-focused therapy: A clinical synthesis. *The Journal of Lifelong Learning in Psychiatry, 3*(1), 32–42.

Greenberg, L. S. (2011). *Theories of psychotherapy. Emotion-focused therapy.* Washington, DC: American Psychological Association.

Greenberg, L. S. (2012). Emotions, the great captains of our lives: Their role in the process of change in psychotherapy. *American Psychologist, 67*(8), 697–707.

Greenberg, L. S., & Angus, L. (2004). The contributions of emotion processes to narrative change in psychotherapy: A dialectical constructivist approach. In L. Angus & J. McLeod (Eds.), *Handbook of narrative psychotherapy* (pp. 331–349). Thousand Oaks, CA: Sage.

Greenberg, L. S., & Pascual-Leone, A. (2006). Emotion in psychotherapy: A practice friendly research review. *Journal of Clinical Psychology, 62* (5), 611–630.

Greenberg, L. S., & Pascual-Leone, J. (1995). A dialectical constructivist approach to experiential change. In R. A. Neimeyer & M. J. Mahoney (Eds.), *Constructivism in psychotherapy* (pp. 169–191). Washington, DC: American Psychological Association.

Hermans, H., & Hermans-Janssen, E. (1995). *Self-narratives: The construction of meaning in psychotherapy.* New York: Guilford Press.

Hermans, H., & Hermans-Konopka, A. (2010). *Dialogical self theory: Positioning and counter-positioning in a globalizing society.* New York: Cambridge University Press.

Jacobsen, B. (2007). What is happiness? *Existential Analysis: Journal of the Society for Existential Analysis, 18*(1), 39–50.

Konopka, A., & Van Beers, W. (2014). Composition work: A method for self-investigation. *Journal of Constructivist Psychology, 27*(3), 194–210.

Lane, R., & Schwartz, G. (1987). Levels of emotional awareness: A cognitive-developmental theory and its application to psychopathology. *American Journal of Psychiatry, 144,* 133–143.

Lyubomirsky, S., King, L., & Diener, E. (2005). The benefits of frequent positive affect: Does happiness lead to success? *Psychological Bulletin, 131,* 803–855.

Mennin, D. S., Heimberg, R. G., Turk, C. L., & Fresco, D. M. (2002). Applying an emotion regulation framework to integrative approaches to generalized anxiety disorder. *Clinical Psychology: Science and Practice, 9,* 85–90.

Rober, P. (2008). The therapist's inner conversation in family therapy practice: Struggling with the complexities of therapeutic encounters with families. *Person-Centered and Experiential Psychotherapies, 7*(4), 245–261.

Robinson, A. L., Dolhanty, J., & Greenberg, L. (2015). Emotion-focused family therapy for eating disorders in children and adolescents. *Clinical Psychology & Psychotherapy, 22*(1), 75–82.

Seikkula, J. (2011). Becoming dialogical: Psychotherapy or a way of life? *Australian and New Zealand Journal of Family Therapy, 32*(3), 179–270.

de Shazer, S., Dolan, Y., Korman, H., McCollum, E., Trepper, T., & Berg, I. K. (2007). *More than miracles: The state of the art of solutions-focused brief therapy* (pp. 37–60). New York: Haworth.

Stith, S. M., Miller, M. S., Boyle, J., Swinton, J., Ratcliffe, G., & McCollum, E. (2012). Making a difference in making miracles: Common roadblocks to miracle question effectiveness. *Journal of Marital and Family Therapy, 38,* 380–393.

Sze, J. A., Gyurak, A., Yuan, Y. W., & Levenson, R. W. (2010). Coherence between emotional experience and physiology: Does body awareness training have an impact? *Emotion, 10,* 803–814.

Honesty and genuine happiness
Or why soft healers make stinking wounds (Dutch proverb)

Nicole Torka

ABSTRACT

Genuine happiness is impossible without authentic concern for and corresponding behaviour towards the well-being of others. Such an incorporation of others into the self refers to a "democratic self" and the related regard for the common good. The author argues that the honesty of professionals who work in or for an educational or vocational setting is vital for the good of the individual and the common good. By introducing "democratic selves", recent advancements in Dialogical Self Theory (DST) point to an inclusion of the common good. However, given the importance of virtues for one's own and the common good, the theory and its applications are in need of integrating virtues and in particular honesty.

Introduction

Many psychologists assume that a happy life is a good life. However, critical scholars remark that those who embrace this thought too often have a blurred idea of genuine happiness and consequently, the good life. Often, there is a lopsided focus on individual happiness and this impedes a look at the importance of morality (see next section). Thus, many psychologists seem to overlook the prominent philosophical idea that an ethical or virtuous life is a necessary (but not sufficient) condition for a genuinely happy or good life. According to Höffe (2007), such a life reconciles individual subjective well-being (*Eigenwohl* or own good) with authentic concern for and corresponding behaviour towards the well-being of others (*Gemeinwohl* or common good). Recently, psychologists Hermans, Konopka, Oosterwegel, and Zomer (2016) advanced Dialogical Self Theory (DST) by implicitly incorporating this idea: Genuine individual well-being requires a "democratic self", a self that takes the voices and needs of others truly into account.

This article contributes to the genuine happiness or good life debate by focusing on virtues and in particular a virtue that affects the own and common good, *honesty*. Even when facing unpleasant facts and the difficult actions these might demand and far-reaching consequences, "an honest person refuses to pretend that facts are other than they are, whether to himself or others" (Smith, 2003, p. 518). I argue that the honesty of professionals who work in or for an educational or vocational setting is vital for individuals own and the common good. After all, the aim of guidance and counselling is growth and those who support others development within the margins of collective customs, needs and practices have to provide truthful information about present issues that might block individuals learning and contribution to community. The next paragraph provides insight into honesty as a virtue and necessary condition for genuine happiness. Then a conceivable force for aligning honesty and genuine happiness will be introduced, Hermans' (2017) Dialogical Self Theory (DST). DST implicitly assumes that honesty in the relationship with others and ourselves is crucial for individual and societal flourishing. As mentioned before, Hermans et al. (2016) advanced DST by introducing the "democratic self", a self that takes the common good into account. According to the authors, for arriving at genuine happiness a truthful (self-)dialogue is inevitable and two questions must be leading: "Do I have sound or realistic beliefs about my needs, aspirations and achievements?" and "Do I really take notice of and, if necessary, adapt my behaviour towards the customs, needs and practices of others?". Given the yet implicit, but central role of honesty in the theory, DST will be discussed in more detail. After this preliminary work it is possible to explore honesty in an educational and vocational setting. The concluding section focuses on the complexity of contemporary societies, associated challenges for supporting one's own and others' democratic selves and the contributing potential of honesty.

Honesty as a virtue and necessary condition for genuine happiness

What motivates humans? Philosophers, religious thinkers, economists (Sen, 1977; Smith, 1759) and sociologists (Parsons, 1937; Polanyi, 1944; Weber, 1922) dispute self-interest as an exclusive catalyst. The homo economicus is challenged and restricted by the *homo moralis* who is motivated by the right thing to do: If the homo moralis is really listening, he or she acts according to, whether or not institutionalised, collective norms and values that are vital for economic, political and social order. Doing the right thing is strongly related to virtues. A virtue is a "disposition or character strength that yields good consequences for the possessor and others" (Hursthouse & Pettigrove, 2016). Several philosophers consider a distinction between "self-regarding" and "other-regarding" virtues as a misconception. Virtues that seem self-regarding, for example, prudence and fortitude are also beneficial for others because a lack can yield negative consequences. Virtues assumed as only important for others such as humanity and justice also benefit the possessor because without them a genuine happy life, *eudaimonia* (Aristotle's Nicomachean Ethics), is impossible (e.g. Hursthouse & Pettigrove, 2016; Taylor & Wolfram, 1968).

Also psychology and in particular moral psychology (e.g. Erikson, 1959; Kohlberg, 1969; Piaget, 1932) is concerned with the homo moralis. However, several scholars criticise the approach and the frequent neglect of morality and virtues in other fields of academic psychology, and especially positive psychology (e.g. Christopher & Hickinbottom, 2008; Han, 2015; Kristjánsson, 2010, 2012; Slife & Richardson, 2008). A liberal individualistic approach towards virtues seems common. If it is at all discussed, psychologists often seem to think of virtues only as a means to an individualistic, self-interest end and neglect virtues meaning for the common good. Moreover, despite the fact that in a clinical setting counsellors' virtues seem to affect client improvement (Natale, 1973; Wilson & Johnson, 2001) and clients' understanding of virtues can contribute to psychological healing (Dueck & Reimer, 2003; Papanek, 1958), Fowers (2005) remarks that applied psychology falls short of virtue discussions.

This article concentrates on one virtue, honesty. "An honest person refuses to pretend that facts are other than they are, whether to himself or others" (Smith, 2003, p. 518). Smith (2003) remarks that we need guidance from others through truthful information, that is the honesty of others, for arriving at sound beliefs about ourselves. Therefore, dishonesty or lying towards others – giving false information, distorting true information or withholding information (Braginsky, 1970) – can be held co-accountable for others inability "to see facts related to the self as they are" (self-honesty). Consequently, without seeing the facts, the person acts, feels and thinks under false impressions. If these impressions are false positives delusional individual well-being might be the consequence. Moreover, dishonesty can also directly affect the possessor by evoking negative emotions like shame, fear of discovery and unhappiness (e.g. Lewis & Saarni, 1993; Ten Brinke & Porter, 2012). Thus, honesty contributes to realistic self-evaluation and reflection of possessors and recipients alike, an ability which is crucial for growth (Dewey, 1933) and is a premise for modifying unhealthy thoughts and behaviours that are a stumbling block to genuine happiness.

Because the aforementioned is based on the ideas of Western scholars, a certain question is legit-imate: Do virtues and in particular the value of honesty differ between and within societies? Based on an extensive literature review of the most influential philosophical and religious traditions (Confu-cianism, Taoism, Buddhism, Hinduism, ancient Greece, Judeo-Christianity and Islam), Peterson and Seligman (2004) report a "surprising amount of similarity across cultures" (p. 36). According to the authors, this indicates the historical and cross-cultural convergence of six core virtues: courage, justice, humanity, temperance, transcendence and wisdom. Honesty (in combination with authen-ticity) is considered as integrity including the strength of courage. Those who study Moral Foun-dations Theory (e.g. Graham et al., 2011; Shweder & Haidt, 1993) have not yet identified honesty as a universal virtue. However, scholars in this field conducted research to determine if honesty, amongst other virtues, should be added as a distinct and universal moral category to the current list of five foundations (i.e. care, fairness, loyalty, authority and sanctity) (Graham et al., 2013).

Dialogical self theory: an aligning force for honesty and genuine happiness?

Before answering this question, a review into the concerns of individualism is inevitable. After all, the preceding sections pointed to objections against a conception of happiness as individualistic well-being: own good separated from common good. According to Dueck and Reimer (2003), a require-ment for clients' understanding of virtues is an awareness of virtue by professionals and a preceding investigation into their own virtue paradigm. However, these authors argue that many psychologists are oblivious of the liberal tradition in the Western psychologists' education and training and that this can be harmful for clients anchored in a different tradition. Moreover, a liberal individualistic ideal is problematic in itself. Liberal individualism has been associated with, for example, a lack of solidarity, exclusion and social injustice (e.g. Fraser, 1997; Honneth, 2011; Taylor, 1979) and Ehrenberg (2004, 2010) argues that this "dictate of individualism" explains increased depression and narcissism in Western individualistic societies (see also Foster, Campbell, & Twenge, 2003).

The well-being of those who suffer from depression and anxiety is far from good. Moreover, because of their frequent expression of antisocial behaviours, narcissists and other dark triad person-alities such as psychopaths and Machiavellians (Paulhus & Williams, 2002) contribute to these con-ditions (e.g. Boddy, 2014; Hare, 1999; Mathieu, Neumann, Hare, & Babiak, 2014). Because of their indifference towards the common good, it could be reasonable to suggest that such egocentric or totalitarian selves (Greenwald, 1980) also suffer. However, several studies show that dark triad person-alities can experience subjective well-being. These results tempt researchers to conclude that such selves may be quite happy (e.g. Egan, Chan, & Shorter, 2014; Rose & Campbell, 2004). In contrast, scholars who observe these disorders longitudinally are sceptical about narcissists' long-term well-being in particular. They need continuous bolstering from others for feeding their self-esteem needs, but due to their often destructive and antisocial behaviour losing positive feedback and social support is very likely (e.g. Fukushima & Hosoe, 2011; Zuckerman & O'Loughlin, 2009). Thus, eventually they harm themselves.

These considerations nourish the idea that genuine happiness is impossible without concern for the common good and Hermans' (2017) Dialogical Self Theory (DST) implicitly supports this con-clusion. Moreover, although psychologists, Hermans et al. (2016) have something important in common with several philosophers (e.g. Hursthouse & Pettigrove, 2016; Taylor & Wolfram, 1968): They remark that a separation between self and others, a mainstream conception in psychology, is a misconception. In Hermans' theory, the self is considered as social and socially constructed. Signifi-cant others, collectives and broader cultures are an intrinsic part of co-organising and constraining the self and others who occupy positions within the "I" contribute to a multivoiced or polyphonic inner dialogue (e.g. Hermans, 2001; Hermans et al., 2016; Hermans & Hermans-Konopka, 2010). "Society is also in the mind" and therefore the theory suggests that the Western ideal of individualism is wishful thinking, because individuals can only function if others play a fundamental role in the self. However, when walking in the footsteps of Ehrenberg (2004, 2010), Hermans (2001) would conclude that if powerful others dictate individualism, these voices can have a dysfunctional effect on the self (e.g. depression, narcissism) and others (e.g. Fraser, 1997; Honneth, 2011; Taylor, 1979).

By discussing a lack of conforming to established practices and customs of collectives and the verbalised judgments of collectives, Hermans (2001) points to others as vital for the moral development and functioning of the self. He acknowledges that others imagined whispers and audible roars co-create, structure, monitor and, if necessary, correct the *homo moralis*. Hermans et al. (2016) imply that in today's world former sharp contours become blurred and this demands more from self's moral authority than in the past. For the own good and the common good, the contemporary democratic self mentally has to cross cultural, gender, national and racial boundaries. Such a self transcends proximate powerful others by also including the voices of distant – alien and less dominant or opposing – groups in the self. Consequently, in the present and future, for the genuine *homo moralis*, listening without pre-judgment to dialogues about challenging ideas and opinions, norms and values is in demand.

Related to norms and values, Hermans and colleagues contrast with scholars who focus on universal virtues. Although Hermans and colleagues do not explicitly discuss virtues in their work, they stress increased diversity; those who focus on universality highlight the historical and cross-cultural convergence of virtues (e.g. Graham et al., 2011; Peterson & Seligman, 2004; Shweder & Haidt, 1993). Listening to the voices of these scholars, it is plausible to assume that we should start to discuss traditionally shared norms and values, which link humans across time and space, before thinking and acting on what drives us apart. In doing so, we can activate the common denominators in our already global mind.

Dialogical Self Theory extends the self to a broad societal context by emphasising the need to listen to distant and alien voices. Moreover, the theory incorporates democracy, a concept that is endemic outside psychology (in philosophy, political science, sociology). In contrast, Valuation Theory and the self-confrontation method (e.g. Hermans & Hermans-Jansen, 1995; Hermans, Fiddelaers, de Groot, & Nauta, 1990; Lyddon, Yowell, & Hermans, 2006), the precursors of DST, are more clearly anchored in social psychology and especially in motivational theories. Moreover, although Valuation Theory and self-confrontation method emphasise that besides personal strength or fulfilled self-interest unity with others is vital for well-being, both understand contact and union with others is a means to achieve the own good. In support of this conclusion, when discussing outcomes with the client, the main focus is on the clients' well-being and implications of and remedies against an unbalanced valuation system; it is not focused on the consequences such imbalances or an undemocratic self can have for others. Finally, in contrast to Valuation Theory, in DST and in particular the democratic self the range of others expands from proximate to include distant and alien others.

In this context, Dialogical Self Theory can be considered as a preliminary, but clear starting point for incorporating the common good into the Valuation Theory and self-confrontation method. By merging own good and common good, theoretical, client-centred and other-centred enhancement towards genuine happiness is possible. A precondition necessary for this endeavour is not yet a subject of the work of Hermans and his colleagues and involves a critical theoretical discussion and incorporation of virtues as well as conscious and purposeful application in counselling. Concerning the reciprocal dialogical relationship between the counsellor and the client in the self-investigation, it can be assumed that for both ideas of one's own and other virtues are essential. Referring to honesty, this means that the counsellor's honesty or truthful information about a valuation and the valuation system is vital for developing and/or supporting clients' ability to see personal meanings and their emotional impact and underlying motives as they are, that is self-honesty (e.g. Papanek, 1958; Putman, 1997; Smith, 2003; Wilson & Johnson, 2001). As mentioned before, honesty of others and self-honesty is crucial for growth and the modification of unhealthy thoughts and behaviours, which can be a stumbling block to genuine happiness.

Honesty in an educational and vocational setting

The development and maintenance of knowledge and skills is essential in an educational and vocational context and professionals such as counsellors, educators and supervisors are vital for the associated learning processes. Dewey (1933) argued that the interference of professionals in such contexts should go further and that they should play an active role in the development of democratic or ethical selves (see also Kristjánsson, 2012). Combining Dewey's idea with assumptions from the Dialogical Self Theory, this means that the efforts of professionals who work in or for such contexts should transcend a contribution to individuals own good – supporting the individual in fulfilling self-motives (self-maintenance and self-expansion) and other-motives (contact and union with others) – by also providing guidance for and monitoring of individuals authentic concern and corresponding behaviour towards others.

Feedback is a powerful force for learning and growth and can be defined as "information provided by an agent regarding aspects of one's performance or understanding" (Hattie & Timperley, 2007, p. 81). Hattie and Timperly's literature review shows that information about task performance and how to become more effective, also termed feedforward, has the highest effect on learning. Lower effects were related to praise, reward and punishment. The authors also mention the positive effects of negative feedback or disconfirmation. In combination with information on how to improve (feedforward), negative feedback can be effective when people have little or no motivation towards a task, they learn new skills or tasks and for stimulating students' enhancement of their own performance goals. However, Hattie and Timperly do not discuss the importance of virtues for performance and understanding-related or other feedback. As mentioned before, in a clinical setting counsellors' virtues seem to affect individuals' improvement. Therefore, it can be assumed that virtues also play an important role in all other environments where improvement like learning and growth is relevant including educational and vocational settings. Natale's (1973) and Wilson and Johnson's (2001) conclusions on virtues important for individual improvement show strong similarities: Natale mentions empathy, genuineness (which includes honesty and openness) and care; Wilson and Johnson describe care, integrity and courage. According to Peterson and Seligman (2004) honesty, in combination with authenticity, means integrity and integrity is a strength of courage.

Why is an honest person courageous? Because the honest person refuses to pretend that facts are other than they are even when facing unpleasant facts, and considering difficult actions that might be demanded and far-reaching consequences (Smith, 2003, p. 518). In a recent article, Ehrmann (2017) discusses grade inflation in German and US education, its negative consequences and the potential sanctions for educational professionals who want to counteract. Research shows that intelligence, knowledge and competences have not increased and possibly have even decreased but grades are higher than in the past. Thus, Ehrmann implies that present grades are less honest and they seem to contain more false positive feedback. Although his focus is on education, it can be assumed that Ehrmann's sketch of consequences is transferable to a vocational context. From a student perspective, investing effort in good grades is less rewarding. Moreover, because many desire high grades, they may opt for less challenging courses. Consequently, individuals' complex knowledge could decrease. Employers' respond to false positive feedback with an increased distrust towards the value of educational institutions evaluations, implement more sophisticated assessment procedures and complain about students' qualities and hubris. Strongly related to hubris, it can be assumed that false positive feedback contributes to a Dunning-Kruger effect or illusory cognitive superiority. Those who suffer from this effect face a double burden: Expertise deficits lead them to make mistakes and these deficits also make them unable to recognise when they are making mistakes (Dunning, 2011, pp. 260–261). Manifestations of hubris often seem to mask task incompetence and are commonly mistaken for leadership potential (Chamorro-Premuzic, 2013). Thus, hubris can catapult the wrong person into a leadership position and false positive feedback can contribute to such a decision error. Without a doubt, selection errors for leadership positions can have a more far-reaching negative impact on the common good than for less powerful positions (Berglas, 2002). Ehrmann refers to another negative effect of unrealistic positive feedback for the common good. Strongly affected by educational institutions grade inflation, employers have increased investment into more sophisticated assessment procedures which widen the inequality gap and therefore undermine democracy. Those with a high socio-economic status can financially invest in preparations for such procedures and therefore can prevail over less wealthy individuals with equal or even better cognitive and social abilities. In other words, compared to moneyed people, people with less means have an added financial obstacle to fulfil their self-motives.

In sum, the contribution of educational institutions to the common good and self-motives related to the own good is at stake. Referring to the own good, false positive feedback from educational and vocational professionals can be held co-accountable for individuals' false beliefs about their performance and therefore fulfilment of self-motives. Overrated students' growth can suffer: After all, they miss a realistic insight into the adequacy of their competences including possible deficiencies and learning needs, which is a stumbling block to genuine happiness. Given the negative consequences for the common good, it is not surprising that Ehrmann advocates revisions and re-strengthening of government interventions in educational policy including protecting those who are and want to be honest about students' achievements. Based on research and common knowledge, he states that educational professionals' honesty about performance is at risk. Honest educational professionals are under pressure from students and parents with the threat of bad assessments and court cases on improving grades. In addition, there is pressure from their employers regarding target agreements for grades and relating grades to faculty financing as well as educators' job security.

Berglas (2002) shows that honest professionals who are focused on union with others (other-motives) and democratic selves can also face dire straits. Although Berglas does not address honesty explicitly and concentrates on executive individuals, his conclusions are partly transferable to others (e.g. students, employees without leadership responsibilities) and point to the challenges honesty contains. Based on his experiences, it can be assumed that the honesty of professionals who work in or for an educational or vocational setting is especially challenged in dealing with individuals with problems. This includes, for example, individuals who suffer from anxiety and/or depression or have a dark triad personality. As mentioned before, the latter can be considered as totalitarian or undemocratic selves and, due to their behaviour, can negatively contribute to the aforementioned debilitating conditions of others.

In contrast to problem individuals who can be trained relatively quickly and painlessly, individuals with problems can best, if at all, be helped by long-term and complicated psychological interventions. Because of their powerful positions and often far-reaching impact on others, according to Berglas, executives with a problem and in particular totalitarian executives are a real and the greatest danger for the common good and claim most from those who intervene. Dark triad personalities are not necessarily incompetent. Regardless of whether they are competent or not, they frequently perceive themselves as good leaders and being high in emotional intelligence (Furnham, Richards, & Paulhus, 2013). They are often also convinced of their spotless functioning. Such a belief in relation to malfunctioning refers to hubris and the related negative consequences described before. However, the moral disengagement or low moral development of such "dark leaders" (Stevens, Deuling, & Armenakis, 2012) can cause severe problems around them including aggressive and deviant behaviour, psychological distress, work-family conflict and a decrease of job performance, satisfaction and organisational commitment (e.g. Boddy, 2014; Mathieu et al., 2014).

Given these potential impacts, it seems clear that timely and adequate intervention is important, but why do individuals with problems demand more from a professionals' honesty than problem individuals? Berglas suggests three reasons. First, those who are negatively affected by malfunctioning individuals demand change as quickly and painlessly as possible. When dealing with individuals with problems, honest professionals cannot promise relatively cheap, rapid and shallow solutions such as behavioural training. As a consequence, the potential client and/or those who pay for interventions might choose a professional who offers an incorrect but relatively inexpensive, fast and superficial solution. Second, professionals who honestly address severe problems can expect ostensible cooperation or severe resistance when the individual is in denial of any personal and/or work-related problems. Finally, individuals with possible deep-rooted problems require self-honest professionals. "Do I possess the right competences and knowledge for guiding this person appropri-ately?" must be the leading question. Despite potentially jeopardising current and future assign-ments, incomes, occupational self-esteem and/or reputation, when the answer is in the negative the professional should make a difficult decision and refer the client to an expert and/or completely withdraw from the case. These actions are also appropriate if the professional concludes that a client cannot be guided, because he or she is not willing or has a very destructive and untreatable form of a dark triad personality. As mentioned before, realistic self-evaluations are not self-evident and Berglas points to the very real dangers of professionals who do not see an underlying deeper problem and overestimate their competences: The psychological problems of the client can be exacerbated and the consequences for his or her surroundings can be disastrous. Thus, professionals who are unable "to see facts related to the self as they are" (self-honesty) and/or are unable or unwilling to provide truthful information to others (honesty towards others) can additionally damage clients' own good and the common good. Although not mentioned by Ehrmann or Berglas, but suggested by Dewey, when professionals legitimately engage to help an individual with problems and in par-ticular if his or her behaviour can pose an imminent or potential future danger for others, they should be "unsolicitedly courageous" by verbalising concerns. This means that professionals have an enlarged responsibility: Their focus has to transcend the own good by also providing guidance and monitoring for the common good.

Conclusions

This article shows that honesty towards ourselves and others is a pre-condition for genuine happi-ness. Without guidance from others through truthful information about ourselves seeing facts related to our self as they are is impossible. Both forms of honesty are crucial for growth and are a premise for modifying unhealthy thoughts and behaviours that are a stumbling block to genuine happiness. Therefore, honesty must be considered as a distinctive and universal virtue. Honesty not only contributes to the own good, the authentic fulfilment of self-motives and other-motives, but also to the common good, the authentic concern for and corresponding behaviour towards the well-being of others, and we have to reconcile both factors for arriving at genuine happiness. However, honesty might be distressing for both the sender and recipient. Therefore, the "how to?" question should be discussed or, in other words, the virtues of honesty. Independent of the context or universally applicable, those who want to contribute to others genuine well-being, learn-ing and growth by providing truthful feedback and information have to take other virtues into account. Based on research about virtues in clinical settings (Natale, 1973; Wilson & Johnson, 2001) and universal virtues (Peterson & Seligman, 2004), it can be assumed that honesty towards others has to be fair, humane – showing care and empathy for the other – and combined with infor-mation on how to adapt or improve thoughts, emotions and behaviour (Hattie & Timperley, 2007).

According to Hermans et al. (2016), currently the common good and the development and con-tinuation of the necessary democratic self, are more difficult to attain than in the past. The world's diversity is more proximate and difficult to ignore. Therefore, contemporary democratic selves, at least mentally, have to cross cultural, gender, national and racial boundaries by truly listening to dia-logues on challenging ideas and opinions, norms and values. This also means that today genuine happiness might be more difficult to achieve than in the past, a past characterised by a more distant world with less information about complex and diverse realities and more opportunities to close one's eyes to these truths. For ensuring the approachability of the precious goal of "genuine happiness" for as many as possible, it is vital to reflect on honesty and other universal virtues that we share with people close to us and those who live on real and symbolical continents apart. The orientation towards this already global part of us, our largely shared virtues, can never start too early and should receive a more important role in clinical settings, education and work life for the sake of democracy and genuine happiness. This is also true for Hermans Dialogical Self Theory and its precursors: Although with respect to democracy and in particular democratic selves, very recent developments are promising (Hermans et al., 2016), the theory is in need of profound consideration and application of virtues.

Disclosure statement

No potential conflict of interest was reported by the author.

References

Berglas, S. (2002). The very real dangers of executive coaching. *Harvard Business Review*. Retrieved from https://hbr.org/2002/06/the-very-real-dangers-of-executive-coaching

Boddy, C. R. (2014). Corporate psychopaths, conflict, employee affective well-being and counterproductive work behaviour. *Journal of Business Ethics, 121*, 107–121. doi:10.1007/s10551-013-1688-0

Braginsky, D. B. (1970). Machiavellianism and manipulative interpersonal behavior in children. *Journal of Experimental Social Psychology, 6*, 77–99. doi:10.1016/0022-1031(70)90077-6

Chamorro-Premuzic, T. (2013). Why do so many incompetent men become leaders? *Harvard Business Review*. Retrieved from https://hbr.org/2013/08/why-do-so-many-incompetent-men

Christopher, J. C., & Hickinbottom, S. (2008). Positive psychology, ethnocentrism, and the disguised ideology of individualism. *Theory & Psychology, 18*, 563–589. doi:10.1177/0959354308093396

Dewey, J. (1933). *How we think. A restatement of the relation of reflective thinking to the educative process* (Revised ed.). Boston: D. C. Heath.

Dueck, A., & Reimer, K. (2003). Retrieving the virtues in psychotherapy. *American Behavioral Scientist, 47*, 427–441. doi:10.1177/0002764203256948

Dunning, D. (2011). The Dunning-Kruger effect: On being ignorant of one's own ignorance. *Advances in Experimental Social Psychology, 44*, 247–296. doi:10.1016/B978-0-12-385522-0.00005-6

Egan, V., Chan, S., & Shorter, G. W. (2014). The Dark Triad, happiness, and subjective well-being. *Personality and Individual Differences, 67*, 17–22. doi:10.1016/j.paid.2014.01.004

Ehrenberg, A. (2004). *La Fatigue d'être soi. Dépression et société*. Paris: Odile Jacob.

Ehrenberg, A. (2010). *La Société du malaise*. Paris: Odile Jacob.

Ehrmann, T. (2017). Liebe Studenten, Sie verdienen schlechte Noten. Retrieved from http://www.zeit.de/campus/2017-05/universitaeten-benotung-studenten-schaden

Erikson, E. (1959). *Identity and the life cycle*. New York: International Universities Press.

Foster, J. D., Campbell, W. K., & Twenge, J. M. (2003). Individual differences in narcissism: Inflated self-views across the lifespan and around the world. *Journal of Research in Personality, 37*, 469–486. doi:10.1016/S0092-6566(03)00026-6

Fowers, B. J. (2005). *Virtue and psychology: Pursuing excellence in ordinary practices*. Washington, DC: APA.

Fraser, N. (1997). *Justice interruptus: Critical reflections on the "postsocialist" condition*. London: Routledge.

Fukushima, O., & Hosoe, T. (2011). Narcissism, variability in self-concept, and well-being. *Journal of Research in Personality, 45*, 568–575. doi:10.1016/j.jrp.2011.07.002

Furnham, A., Richards, S. C., & Paulhus, D. L. (2013). The dark triad of personality: A 10 year review. *Social and Personality Psychology Compass, 7*, 199–216. doi:10.1111/spc3.12018

Graham, J., Haidt, J., Koleva, S., Motyl, M., Iyer, R., Wojcik, S. P., & Ditto, P. H. (2013). Chapter two – Moral foundations theory: The pragmatic validity of moral pluralism. *Advances in Experimental Social Psychology, 47*, 55–130. Retrieved from SSRN: https://ssrn.com/abstract=2184440

Graham, J., Nosek, B. A., Haidt, J., Iyer, R., Koleva, S., & Ditto, P. H. (2011). Mapping moral domain. *Journal of Personality and Social Psychology, 101*, 366–385. doi:10.1037/a0021847

Greenwald, A. G. (1980). The totalitarian ego: Fabrication and revision of personal history. *American Psychologist, 35*, 603–618. doi:10.1037/0003-066X.35.7.603

Han, H. (2015). Virtue ethics, positive psychology, and a new model of science and engineering ethics education. *Science and Engineering Ethics, 21*, 441–460. doi:10.1007/s11948-014-9539-7

Hare, R. D. (1999). *Without conscience: The disturbing world of psychopaths among us*. New York: The Guilford Press.

Hattie, J., & Timperley, H. (2007). The power of feedback. *Review of Educational Research, 77,* 81–112. doi:10.3102/003465430298487

Hermans, H. J. M. (2001). The dialogical self: Toward a theory of personal and cultural positioning. *Culture & Psychology, 7,* 243–281. doi:10.1177/1354067X0173001

Hermans, H. J. M. (2017). *Research program on "Valuation theory, self-confrontation and dialogical self".* Retrieved from http://huberthermans.com/research-program-valuation-theory-self-confrontation-and-dialogical-self/

Hermans, H. J. M., Fiddelaers, R., de Groot, R., & Nauta, J. F. (1990). Self-confrontation as a method for assessment and intervention in counseling. *Journal of Counseling & Development, 69,* 156–162. doi:10.1002/j.1556-6676.1990.tb01478.x

Hermans, H. J. M., & Hermans-Jansen, E. (1995). *Self-narratives: The construction of meaning in psychotherapy.* New York: Guilford Press.

Hermans, H. J. M., & Hermans-Konopka, A. (2010). *Dialogical self theory: Positioning and counter-positioning in a globalizing society.* Cambridge: Cambridge University Press.

Hermans, H. J. M., Konopka, A., Oosterwegel, A., & Zomer, P. (2016). Fields of tension in a boundary-crossing world: Towards a democratic organization of the self. *Integrative Psychological and Behavioral Science,* in press. doi:10.1007/s12124-016-9370-6

Höffe, O. (2007). *Lebenskunst oder Moral: oder macht Tugend glücklich?* München: C.H. Beck.

Honneth, A. (2011). *Das Recht der Freiheit – Grundriß einer demokratischen Sittlichkeit.* Frankfurt a. M.: Suhrkamp.

Hursthouse, R., & Pettigrove, G. (2016). Virtue ethics. *The Stanford Encyclopedia of Philosophy,* Winter Edition. Retrieved from https://plato.stanford.edu/archives/win2016/entries/ethics-virtue/

Kohlberg, L. (1969). Stage and sequence: The cognitive development approach to socialization. In D. A. Goslin (Ed.), *Handbook of socialization theory and research* (pp. 347–480). Chicago, IL: Rand McNally.

Kristjánsson, K. (2010). Positive psychology, happiness, and virtue: The troublesome conceptual issues. *Review of General Psychology, 14,* 296–310. doi:10.1037/a0020781

Kristjánsson, K. (2012). Virtue development and psychology's fear of normativity. *Journal of Theoretical and Philosophical Psychology, 32,* 103–118. doi:10.1037/a0026453

Lewis, M., & Saarni, C. (1993). *Lying and deception in everyday life.* New York: The Guilford Press.

Lyddon, W. J., Yowell, D. R., & Hermans, H. J. M. (2006). The self-confrontation method: Theory, research, and practical utility. *Counselling Psychology Quarterly, 19,* 27–43. doi:10.1080/09515070600589719

Mathieu, C., Neumann, C. S., Hare, R. D., & Babiak, P. (2014). A dark side of leadership: Corporate psychopathy and its influence on employee well-being and job-satisfaction. *Personality and Individual Differences, 59,* 83–88. doi:10.1016/j.paid.2013.11.010

Natale, S. M. (1973). Interpersonal counsellor qualities: Their effect on client improvement. *British Journal of Guidance and Counselling, 1,* 59–65. doi:10.1080/03069887308259352

Papanek, H. (1958). Ethical values in psychotherapy. *Journal of Individual Psychology, 14,* 160–166. Retrieved from https://search.proquest.com/openview/8f02f9a1971bcd339d81a2bc481656f4/1?pq-origsite=gscholar&cbl=1816607

Parsons, T. (1937). *The structure of social action.* New York: McGraw Hill.

Paulhus, D. L., & Williams, K. M. (2002). The dark triad of personality: Narcissism, Machiavellianism, and psychopathy. *Journal of Research in Personality, 36,* 556–563. doi:10.1016/S0092-6566(02)00505-6

Peterson, C., & Seligman, M. E. (2004). *Character strengths and virtues: A handbook and classification.* Washington, New York: American Psychological Association, Oxford University Press.

Piaget, J. (1932). *The moral judgment of the child.* New York: The Free Press.

Polanyi, K. (1944). *The great transformation: The political and economic origins of our time.* Boston: Beacon Press.

Putman, D. (1997). Psychological courage. *Philosophy, Psychiatry, & Psychology, 4,* 1–11. doi:10.1353/ppp.1997.0008

Rose, P., & Campbell, K. W. (2004). Greatness feels good: A telic model of narcissism and subjective well-being. In S. P. Shohov (Ed.), *Advances in psychology research* (Vol. 31, pp. 3–26). New York: Nova Science.

Sen, A. K. (1977). Rational fools: A critique of the behavioral foundations of economic theory. *Philosophy & Public Affairs, 6,* 317–344. Retrieved from http://www.jstor.org/stable/2264946?seq=1#page_scan_tab_contents

Shweder, R., & Haidt, J. (1993). Commentary to feature review: The future of moral psychology: Truth, intuition, and the pluralist Way. *Psychological Science, 4,* 360–365. doi:10.1111/j.1467-9280.1993.tb00582.x

Slife, B. D., & Richardson, F. C. (2008). Problematic ontological underpinnings of positive psychology. *Theory & Psychology, 18,* 699–723. doi:10.1177/0959354308093403

Smith, A. (1759). *The theory of moral sentiments.* New York: Penguin. (2010).

Smith, T. (2003). The metaphysical case for honesty. *The Journal of Value Inquiry, 37,* 517–531. doi:10.1023/B:INQU.0000019033.95049.1e

Stevens, G. W., Deuling, J. K., & Armenakis, A. A. (2012). Succesful psychopaths: Are they unethical decision-makers and why? *Journal of Business Ethics, 105,* 139–149. doi:10.1007/s10551-011-0963-1

Taylor, C. (1979). Atomism. In A. Kontos (Ed.), *Powers, possessions and freedoms: Essays in honor of C. B. Macpherson* (pp. 39–61). Toronto: University of Toronto Press.

Taylor, G., & Wolfram, S. (1968). The self-regarding and other-regarding virtues. *The Philosophical Quarterly, 18,* 238–248. doi:10.2307/2218561

Ten Brinke, L., & Porter, S. (2012). Cry me a river: Identifying the behavioral consequences of extremely high-stakes inter-personal deception. *Law and Human Behavior, 36*, 469–477. doi:10.1037/h0093929

Weber, M. (1922). *Wirtschaft und Gesellschaft*. Tübingen: J. C. B. Mohr.

Wilson, P. F., & Johnson, W. B. (2001). Core virtues for the practice of mentoring. *Journal of Psychology and Theology, 29*, 121–130. Retrieved from https://search.proquest.com/openview/3b231819740657ca340f3815a865fab6/1?pq-origsite=gscholar&cbl=47846

Zuckerman, M., & O'Loughlin, R. E. (2009). Narcissism and well-being: A longitudinal perspective. *European Journal of Social Psychology, 39*, 957–972. doi:10.1002/ejsp.594

Functions of internal temporal dialogues

Małgorzata Łysiak and Małgorzata Puchalska-Wasyl

ABSTRACT

Psychological literature in the field of internal dialogical activity assumes that internal temporal dialogues perform several important functions, namely: support, redefining the past, balancing, distancing, advising, making decisions, acquiring wisdom and managing the future. The article is an attempt to verify this proposal through qualitative analyses of temporal dialogues conducted by three persons, who participate in the wider research project on functions of internal temporal dialogues ($N = 200$). Dialogical Temporal Chair Technique was used. Presented qualitative analyses of internal temporal dialogues seem to confirm most of the functions listed in the theoretical proposition. The results are also discussed with reference to well-being and happiness as well as the theory of the dialogical self (the role of metaposition) and time perspective.

Travelling in time – the eternal human desire. Easy time travel seems like a tempting prospect that could bring a lot of profits. Charles Dickens in his "A Christmas Carol" has shown a profound transformation of Ebenezer Scrooge – a man who on Christmas Eve could return to his past thanks to ghosts, and then see his future and in effect make a thorough reflection on his life. It is just fiction. But are the positive effects of travelling in time indicated by Dickens – distancing oneself from the current situation by looking at it from a past or future perspective, or acquiring wisdom expressed in the discovery of some truth about life, the importance of interpersonal relations or values – actually the effects unattainable for a human who has not invented the time machine yet?

The French phenomenalist Maurice Merleau-Ponty, referring to the concept of an intentional arc, decades ago wrote about an extraordinary quality of the human mind that allows a person to travel in time and refer to themselves from the perspective of a specific point in the future or the past (cf. Hermans & Kempen, 1993). Referring to this ability, Hermans (1996, p. 33) states:

> I can imaginatively move to a future point in time and then speak to myself about the sense of what I am doing now in my present situation. This position, at some point in the future, may be very helpful to evaluate my present activities from a long-term perspective. The result may be that I disagree with my present self as blinding itself from more essential things.

Such an assessment of the current activity may be a consequence of a change of perspective and a simple comparison of the results of the adoption of two different evaluation perspectives. Sometimes, however, such a comparison takes the form of an internal dialogue, that is, a person alternates (at least) two points of view and the statements formulated from these perspectives respond to one another (Puchalska-Wasyl, 2015; cf. Hermans, 2003; Hermans & Hermans-Jansen, 1995).

The concept of internal dialogue fits into a wider theoretical context defined by such concepts as: self-talk, private speech, inner speech, egocentric speech or internal monologue (cf. Depape,

Hakim-Larson, Voelker, Page, & Jackson, 2006). However, if other terms suggest that the speaker and the receiver of the statements are the same, then the "internal dialogue" assumes that there are at least two communicating parties within one person. Thus, there is the idea reflected here of a non-monolithic self, widely accepted in psychology – in polipsychism (Assagioli, 2000) and in the cognitive psychology of the self (Higgins, 1987; Markus & Nurius, 1986), but above all in the dialogical approach (Hermans, 2003; Hermans & Hermans-Jansen, 1995).

The dialogical approach (Hermans & Gieser, 2012), on the basis of which the notion of internal dialogue emerged, assumes that a person can take many different viewpoints (perspectives), here referred to as I-positions. Dialogical self in the light of the Hubert Hermans's idea is a dynamic multiplicity of such relatively autonomous I-positions, each of which is endowed with its own voice. This means that each of them is able to give expression to specific beliefs, feelings and motivations significant for a given point of view, create a narrative around it as well as enter into a dialogue with another I-position. I-positions are shaped in different social contexts and may represent, for example, culture, community or a significant person. Therefore, a person can consider a problem from the point of view of the group to which they belong, but also from a personal perspective. I-positions may also express some aspect of themselves, felt as important and separate in relation to other aspects of themselves or they may represent a given person at different moments of their life. As a result, it happens that our "good self" argues with the "bad self" on the moral issue, or the "enthusiastic self" tries to encourage the "passive self" to act. It also happens sometimes that I in the future encourages I in the present or I in the present accuses of neglecting I in the past.

In a situation where the internal dialogue concerns the exchange of views between the voices representing two different and distant in time points of view, we are talking about temporal dialogues. Temporal dialogues usually occur between I in the past and I in the present or between I in the future and I in the present, much less often between I in the future and I in the past. It turns out that such a dialogical exchange of "voices" from the past or the future can have a particular impact. The research presented here focuses on such temporal dialogues.

Sobol-Kwapińska and Oleś (2010) talk about some basic functions of temporal dialogues, which are as follows: support, redefining the past, balancing, distancing, advising, making decisions, acquiring wisdom and managing the future.

Support results from the fact that both I in the future and I in the past may act as a comforter in difficult moments and send various types of supporting messages, for example, "You can do it, you'll see. You've done it many times before, so now it will work, too!". Redefining the past is associated with an attempt to look at past events in a new way and negotiate a new meaning for them. The result may be, for example, working through a negative experience and freeing oneself from ruminations. Balancing involves assessing the value of actions taken in the past, as well as evaluating profits and losses from a perspective in which their effects are already known. Balancing is also trying to determine what could be done differently and what not. As the authors emphasise, the balance made in a dialogue form protects against a unilateral and thus skewed assessment. Distancing is a detachment from a person's current experience and an attempt to look at it from the perspective of the past or the future, allowing them to evaluate it in a new way. As a result, a person realises that hard times (just like significant successes) will end sometime. Advising consists in formulating advice and guidance from the perspective of past experiences or anticipated future states. I in the future, as well as I in the past, may be treated by a person as their life advisor. Especially contact with the latter aspect of the self – I in the past – seems to facilitate the use of coping resources that once proved to be effective. Help in making decisions results from confronting the reasons coming from different temporal perspectives, which is particularly useful when making critical life decisions or choosing life goals. A person in such a dialogue can put themselves in a situation of two alternative possibilities or take the perspective of the ultimate balance of life. Acquiring wisdom is a consequence of using the position of I in the future and/or the I in the past to capture a wisdom perspective, formulate some truth about life, a maxim about its meaning, the importance of interpersonal relations or values. Finally, managing the future is a function of planning changes and preparing for future challenges by creating a possible self from the imagined future (Markus & Nurius, 1986). Creating a specific possible self, especially when having a conversational contact with it, on the one hand, prevents the temptation of dreaming unreal dreams, and on the other hand, allows to embrace the future and make it more specific by formulating goals and plans, as well as evaluate the purpose of these goals, which strengthens or weakens the motivation to achieve them.

In the outlined context, internal temporal dialogues seem to fulfill very important functions in the process of learning and conscious shaping of a person and their future, which can be crucial for people standing at the threshold of adulthood – adolescents and young adults. However, Sobol-Kwapińska and Oleś (2010) when discussing the above functions, stress that this is only a theoretical proposal, requiring empirical confirmation. The research presented below meets the need signalled by these authors. The purpose of the qualitative analyses is to seek answers to the following questions:

(1) What are the main topics of internal temporal dialogues in adolescents and people in early adulthood?
(2) What functions can internal temporal dialogues perform in these age groups?

Method

Participants

Participants of the study were people at two different development stages: 100 people during adolescence, including 56 women ($M_{age} = 17.92$; SD = 1.01) and 100 people during early adulthood, including 60 women ($M_{age} = 22.96$; SD = 2.38). The adolescence is a time of change and the first life balance when the development of identity is the main task. It is a time when new I positions emerge because the individual often tries to answer many difficult questions while looking for their identity. Young adulthood is a moment for searching goals, new tasks and making the sense of life, but nowadays it seems that also is a moment where the identity issues are still actual. The young adults seem to ask themselves the same questions as teenagers and they are more prone to stay in their role of teenager (e.g. staying with parents, lengthen their studying). Because the process of conscious shaping of identity and personal future appears to be especially connected with internal temporal dialogues, both these groups were analysed in our study.

The selection of the research group was not random. In the group of adolescents, there was a representation of secondary school students (51 people) and vocational school students (49 people), who were invited to the study by ads posted in their schools, during lectures, from periods and through cooperation with school educators. The young adults were people studying humanities (38 people), technical subjects (34 people), as well as legal and economic subjects (28 people). The respondents joint the research via advertisements that had been distributed at universities, dormitories or student meeting places. All the respondents participated in a wider research project on changing the meaning of life under the influence of temporal dialogues.

Procedure

The study lasted about 60 minutes. The dialogue was triggered by the Dialogical Temporal Chair Technique (DTCT) constructed by Łysiak (2017; Łysiak & Oleś, 2017). The instruction was inspired by the techniques used in therapeutic Gestalt and cognitive–behavioural approaches. There were three chairs in front of the subjects, each of them symbolising a self in a given time: in the middle – I in the present, on the left – I in the past, on the right – I in the future. A person who was describing an important moment from their past sat on a chair symbolising the past. They recalled past events and situations with all details possible, such as emotional climate or psychosocial context. Then they directed a message to their I in the present, which responded back so that the exchange could take place. At the end of the dialogue procedure, a question was asked: What was the result of juxtaposing these two voices, can they be combined into a common message? In this way, it was checked whether there is a reflection coming from a given dialogue, a bridge by which a person is able to connect two temporal dimensions with each other. In a similar way a dialogue between I in the future and I in the present was activated. A participant was asked to imagine what their life would look like in 10 years' time and describe how they see themselves, what they are, what they do, in short, they were asked to try to empathise with themselves in the future. Then they addressed a message to their I in the present, which responded back. The dialogue was followed by a reflection linking standpoints of both temporal positions. The subjects were completely free to choose "selves" from the past and to imagine "selves" in the future. In principle, the method did not impose the topics of the dialogues. At the end of the whole procedure of activating the voices, the respondents were asked for a metareflection, that is a look at their two dialogues with a metaperspective that is characteristic of the so-called metaposition (Hermans, 1996). Metareflection was aimed at compiling, combining and interpreting the voices flowing from the temporal I-positions, taking into account the time of events and the intertwined threads (Hermans, 2001, 2003). This metareflection, a new quality, was expressed by the participants in a summary sentence, a metaphor or a message. The whole procedure including two dialogical exchanges (I in the past vs I in the present; and I in the future vs. I in the present) as well as metareflection, was schematically illustrated in Figure 1. The course of the dialogues was recorded on an ongoing basis (with the consent of the respondents) and documented in writing by the researcher.

Results

The content analysis of all the collected dialogues was carried out ($N = 200$) in the search for the answer to question 1. The written dialogues in random order were shown to the four competent judges. In the first stage, each judge taking into account the content of the internal temporal dialogues, distinguished several main topics. In the second stage, the judges jointly compared the categories and discussed the discrepancies in order to eliminate them. The effect of their work is presented in Table 1.

One of the most common topics of the dialogues were achievements in the area of the education and school career. In this respect, references have been made to both successes and failures. In these types of dialogues, young people often analysed their life choices in terms of success, so as to confirm their decisions. I in the future became the desirable possible self (Markus & Nurius, 1986), which served to build motivation so that a person could realise their goal in the future. The participants confronting temporal I-positions also did not avoid topics related to failure, which was accompanied by a rather negative affective climate.

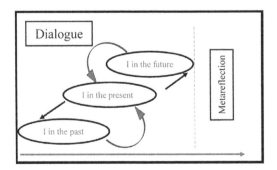

Figure 1. "Dialogical Temporal Chair Technique" – a schematic diagram of a dialogue procedure for the activation of the voices (own elaboration, source: British Journal of Guidance and Counselling, 2017).

Table 1. Topics of internal temporal dialogues.

	Participants		
Topics of dialogues	Total (N = 200)	Students (N = 100)	Teenagers (N = 100)
Career, education (studies, work, school)	34%	31%	37%
Family	27.5%	27%	28%
Love, feelings	21%	27%	15%
Sense of life	11%	8%	14%
Time	5.5%	7%	4%
Trauma (death of relatives, violence)	1%	–	2%

Another topic discussed in the dialogues was the family. In the position of I in the past, the participants returned to their experiences with their parents, and when talking about the future, they saw visions of their own families, often different from the one from the past. Emotivity varied depending on the adopted temporal perspective. I in the future seldom projected visions of an unhappy family, a wrong choice of a partner or a failed relationship. However, there were references to unpleasant experiences with both parents and siblings in the descriptions of I in the past.

Another topic that appeared in the dialogues was the subject of feelings, especially love. It evoked specific life events and emotions associated with them. The affective effect of some dialogues was rather negative than positive. The subject of love was expressed not only in the form of memories or imaginations about happiness, joy and motivation to live, but also implied suffering and referred to the existential dilemmas of a person, such as love and loneliness. For example: Can I change for love? Can I love since I am what I am? Jacobsen (2007), writing about existential dilemmas, understands them as situations in which a person finds themselves between two polarised poles and has to make a choice, but no option is optimal. Dilemmas, which are mentioned by the author, include happiness-suffering, love-loneliness, and sense-senselessness of life. According to Jacobsen, these dilemmas are interconnected, so if a person suddenly experiences a death of a loved one, a deep suffering, they simultaneously experience loneliness and the meaninglessness of life. It seems that in temporal dialogues there are contradictions similar to existential dilemmas. Confronting the experience of I in the past, focused on unpleasant situations and negative emotions with supporting I in the present, a person can simultaneously experience suffering and satisfaction, sadness and happiness. The paradox lies in the fact that satisfaction of I in the present is often a result of what a person has learned from a difficult past experience.

Several dialogues referred directly to the meaning of life, existence and its purpose, and time frames. Confronting who I am and who I will be among the youth, as well as the balance of the past achievements in early adulthood are the threads that in themselves reflect the value and sense of life.

One of the teenagers recalled an extremely painful situation in her dialogue, namely her father's death. Despite the fact that this dialogue was about family, due to its affective climate and its special expression in the context of other dialogues, judges classified it into a separate thematic category including traumatic events.

Below, in the search for the answers to question 2, a qualitative content analysis of three dialogues was carried out. The first of these will be the aforementioned dialogue on father's death by 19-year-old Katarzyna.

1. **DESCRIPTION OF I IN THE PAST:** *One of the key events that changed my whole life, perception of the world, values was my dad's death. I'm 13 years old. I'm finishing primary school, I have friends, everything is going well and I'm doing fine. And he suddenly does not return home from work. Initially, we think with my mom that he has overtime, so that's ok. After a few hours, a phone call, this terrible phone call. Dad had a heart attack. The first and the last. We're going to the hospital. Nothing is the same after that.*

I IN THE PRESENT: *I still can't deal with it. It's been so many years. Death changed everything.*

I IN THE PAST: *The worst thing in all of this was returning to reality, ordinary activities, breakfasts, dinners. Just the two of us. And that damned sadness.*

I IN THE PRESENT: *Paradoxically, it strengthened you. Would you change something then? Would you do something differently?*

I IN THE PAST: *Probably not. I won't turn back time, and the worst thing is that I can't find the answer: why?*

I IN THE PRESENT: *Sometimes it's better not to know the answer. I still don't know why it happened, but I do know that you did well then. You were a huge support for your mom.*

I IN THE PAST: *Yes. I know. Sometimes I dream about seeing him, make him smile again, or even shout at me. Just being present.*

REFLECTION: *Time can be terribly fickle. Sometimes it heals, sometimes it reopens old wounds.*

2. DESCRIPTION OF I IN THE FUTURE: *I am a young teacher. I feel professionally fulfilled. I have my own flat, a loving husband and a child. I try to really take care of myself and my family, especially in the context of health. Sometimes I return to the past and regret that my dad can't see my happiness.*

I IN THE PRESENT: *Looking at your past life, I'm glad that you succeeded. That I think you are happy after all.*

I IN THE FUTURE: *I am and I know that you will do everything to make me happy. I think it's my-your misfortune that has strengthened us.*

I IN THE PRESENT: *I know, although sometimes it's still not easy for me, but from what you're saying it will heal somehow and although memory does not go away, life can be beautiful.*

I IN THE FUTURE: *It can be, and will be. And if you're doing well, it would make our dad proud. He's somewhere and watches over us, I know it.*

REFLECTION: *Thanks to what I see in the future, I can better fulfil myself.*

METAREFLECTION: *Time is a connector. Time allows me to remember and at the same time plan a beautiful life.*

As part of a narrative approach in psychology, there are types of analysis of life stories of people who have experienced painful situations in their lives, such as the loss of a loved one. Such events, especially unexpected ones, are difficult to assimilate with past experiences as well as expectations regarding the future (Neimeyer, 2006). The discussed dialogue made it possible to **redefine past experience**. Sadness, despair and hopelessness typical for I in the past (*I won't turn back time, and the worst thing is that I can't find the answer: why?*) are replaced by more positive states. Despite the experience of loss there is a shadow of hope, understanding and looking into the future in the dialogue between I in the present (*(…)from what you're saying it will heal somehow and although memory does not go away, life can be beautiful*) and I in the future (*It can be, and will be*). One can get an impression that sharing this difficult experience with themselves has a function of **balancing**, because the cognitive reworking of a dramatic event makes the person sure that it was impossible to do anything more (I in the present: *Would you change something then? Would you do something differently?*; I in the past: *Probably not*). In addition, the support given to a particular I-position restores the person's sense of control and probably leads to a rise in well-being (Reflection: *Thanks to what I see in the future, I can better fulfil myself*). Consequently, it may have a positive connection to the elements of posttraumatic growth seen in the dialogue. Ultimately, the girl expresses the hope that she will be happy and tries to plan a beautiful life (Metareflection: *Time allows me to remember and at the same time plan a beautiful life*).

Internal dialogues help to organise incomprehensible, difficult, sometimes chaotic experiences. Looking at the temporal part of the self from a metalevel allows a person to distance themselves from their experiences, thanks to which the person has a chance to get to know and understand themselves better. An example of this is a dialogue by Filip, a 24-year-old student of philosophy.

1. DESCRIPTION OF I IN THE PAST: *The moment I often recall is a time in secondary school, I'm about 17 years old and I enjoy reading Nietzsche. I'm fascinated by his philosophy and his life. I am looking for his biography, I spend time in libraries reading his books. Generally, I don't notice that I am beginning to take over his views, I move away from people, I start to live as if there is nothing afterwards. This decadence of mine means for me being someone, someone beyond what is happening around me. No, I'm not going into fascism, I rather come to the conclusion that life does not make much sense.*

    I IN THE PRESENT: *You are a young boy, how can you say that life does not make sense? You have your life ahead of you.*

    I IN THE PAST: *I do not know that now. For me, life's about a book and being a loner. I see no point in parties, girls, arguing with teachers, as my peers do. On the one hand, I'm above it, on the other, I feel like I'm a freak.*

    I IN THE PRESENT: *Because you are a freak. You shut yourself off, you don't talk to anyone, you look like a freak. Only, where is so much pessimism in you coming from?*

    I IN THE PAST: *Because I don't see too much point, I prefer to surround myself with the culture of meaninglessness, worthlessness, reflection without reflection in a sense. I do not know what will happen to me in a few years, and that's how I feel at the moment.*

    REFLECTION: *Finding the meaning in life requires effort, it is probably worth striving for it.*

2. DESCRIPTION OF I IN THE FUTURE: *It's going to be difficult to describe. I don't know who I'm going to be in 10 years' time. I'd really like, and I'm going this path to be someone known and valued, mainly in the world of science. Why? Because then I have the chance to leave some legacy, find a deep sense of my actions. I think that life is going to get more exciting and I'll be much more motivated in 10 years' time. I want to do things.*

    I IN THE PRESENT: *I think that your philosophy of life is changing. You already perceive the sense of your actions and do everything to devote yourself to the scientific career.*

    I IN THE FUTURE: *Yes, but your motivation is unstable. Sometimes you go forward, you are even successful, but one failure and you say that life doesn't make sense. Maybe it is worth going a little bit towards real expectations, not only imagined ones and give vent to what is important for you? Life is beautiful, as Benigni said.*

    I IN THE PRESENT: *Maybe it is beautiful, but if you can't see this beauty for many years, then it's not so simple to do it. Sometimes it seems to me like, generally, the meaning is impossible to find.*

    I IN THE FUTURE: *And I think that it's possible and that you will find it, too, despite the elusiveness of life, despite the delicacy and fragility and often senseless events that affect you. It's important to notice what you sometimes deliberately avoid and it's easier for you to say that everything is meaningless.*

    REFLECTION: *I want to find the meaning, although I have doubts whether I can do it.*

    METAREFLECTION: *It is worth looking for the meaning of life, even if sometimes I have to undermine many of my decisions.*

This is an example of a dialogue in which a speaker is negotiating with himself the value of his life and the meaning of his actions. It is typical that temporal positions sometimes speak using different voices, seemingly contradictory. For example, on the one hand, they support what the speaker is saying (I in the present to I in the future: *Sometimes it seems to me like, generally, the meaning is impossible to find*), and on the other hand, they are as if outraged by the speaker's attitude (I in the present to I in the past: *(…) how can you say that life does not make sense? (…) Because you are a freak*). It is worth noting that the author of this dialogue, describing his I in the future, also seems to be conducting a dialogue within this position (*I'd really like, and I'm going this path to be someone known and valued, mainly in the world of science. Why? Because then (…)*). This dialogue illustrates what Hermans (1996) calls the multiplicity of voices – it is important which of them comes to the foreground and what consequences it will bring. The positions of I in the past, I in the present and I in the future are saturated with a specific affective climate. I in the past has negative experiences that Filip would like to change (*For me, life's about a book and being a loner. I see no point (…). I feel like I'm a freak*), I in the future expresses hopes and plans, goals that the young man wants to achieve (*I'd really like, and I'm going this path to be someone known and valued, mainly in the world of science. (…) I want to do things*). **The position of I in the present** seems to be **mobile and multi-functional** depending on with which temporal position the dialogue is being led. In the conversation with I in the past, I in the present becomes an understanding friend and mentor, providing emotional support (*You have your life ahead of you*), encouraging and helping I in the past to find a distance from past events, which is expressed in the Filip's reflection: *Finding the meaning in life requires effort, it is probably worth striving for it.* However, in the confrontation with I in the future, I in the present becomes the one who needs advice, support and sometimes consolation (*Sometimes it seems to me like, generally, the meaning is impossible to find*). In this case, I in the future seems to take over functions that I in the present has just had, becoming a mentor and an understanding friend (*And I think that it's possible and that you will find it, too, despite the elusiveness of life (…). It's important to notice what you sometimes deliberately avoid and it's easier for you to say that everything is mean-ingless*). On the basis of Hermans' theory, the I-positions and its functions change in space and time taking on a different voice, and roles in conversations between one another.

The referenced dialogue illustrates existential dilemmas (Jacobsen, 2007). This young boy is won-dering about the meaning of life and its meaninglessness, at the same time the topic of the basic dilemma appears, whether to be himself and with himself or with other people. During the dialogue, there is no direct solution, however, there is the impression that I in the future is trying to give direc-tions to resolve the internal dispute. Confronting voices from distant temporal positions and identify-ing with them allows a person to evaluate current events using a different perspective. Allowing various contents of the voices to be adopted gives meaning to the present dilemmas experienced here and now. The exploration of the voices confirms the existing beliefs, however, it also hints their change. The young man has a chance to better understand who he was, who he is, and who he would like to be and if and how it is possible. In addition to the feeling of a meaningless life that is rooted in his past history, he finds a strong desire to leave his mark in the field of science some-time in the future. He confronts the future vision with the current state to determine to what extent it is possible to achieve. As I in the present he notices: *You already perceive the sense of your actions and do everything to devote yourself to the scientific career*, but in one moment from another temporal pos-ition, as I in the future, he adds that he is indecisive and that his expectations should be more realistic (*Yes, but your motivation is unstable. Sometimes you go forward, you are even successful, but one failure and you say that life doesn't make sense. Maybe it is worth going a little bit towards real expectations, not only imagined ones and give vent to what is important for you?*). Thus, the presented dialogue is an attempt to get to know oneself, and also a step towards a more conscious creation of oneself.

Referring to the functions proposed by Sobol-Kwapińska and Oleś (2010), the presented dialogue seems to be primarily a source of **support** and **advice**. I in the future fulfils the role of a mentor that shares their knowledge and tries to convey their wisdom to I in the present. In this sense, dialogue also favours **acquiring wisdom**.

The dialogue by 23-year-old Anna allows to see other functions of internal dialogue between tem-poral positions.

1. DESCRIPTION OF I IN THE PAST: *I can't make a decision. I've always been like this. I didn't know which school to choose, I didn't have anyone to ask advice for. It was the same with my interests, everyone had something and I could do everything and nothing. I couldn't choose a boyfriend, either, and always ended up alone at a party. I remember that going to the university was similar. I was always think-ing about tomorrow, what would happen, who would I be, but I could not decide.*

I IN THE PRESENT: *You didn't use all the opportunities because you're not patient, you wanted to have everything at once, and it doesn't work like that. You will slowly learn how to choose between what you should do and what you really want to do.*

I IN THE PAST: *Do you believe that I can change? That in the end the world will be in my favour?*

I IN THE PRESENT: *I know that in a few years you will look at the world around you a little differently and you will notice your indecision or even lack of determination as an advantage, not a flaw and reasons to be sad.*

I IN THE PAST: *I want to believe that what you say is true. It's always worth trying.*

REFLECTION: *I am glad of my experiences. Thanks to this, it's easier for me now.*

2. DESCRIPTION OF I IN THE FUTURE: *I've got what I wanted. Professional success, I am fulfilling myself, I feel happy going to work and satisfied with the actions taken. It cost me a lot, but the most important thing was not to give up and have hope. I'm happy.*

I IN THE PRESENT: *Everything's not as easy as it looks from a different time perspective.*

I IN THE FUTURE: *Oh, you just want to justify yourself. The easiest way is to say that life is not easy, instead of taking matters into your own hands.*

I IN THE PRESENT: *You're smarter, but you really have what you want. So I have a chance to change something despite what has happened.*

I IN THE FUTURE: *It's normal that you have doubts whether it'll work, but it's worth looking at the future and not worrying about what has been, just act.*

REFLECTION: *Seeing yourself in retrospect makes it easier to not worry about what is happening now.*

METAREFLECTION: *In retrospect, everything seems simple. It's going to work. I'm happy that I can live in such a way that I still have this perspective ahead of me.*

The analysis of Anna's dialogue suggests that the dialogue performs several important functions. In the confrontation of I in the past with I in the present, I in the present supports, and simultaneously persuades I in the past to look differently at what is to come (*You didn't use all the opportunities because you're not patient, you wanted to have everything at once, and it doesn't work like that. You will slowly learn how to choose between what you should do and what you really want to do (…) I know that in a few years you will look at the world around you a little differently (…).* Looking at the reflection after the dialogue between these two temporal positions (*I am glad of my experiences. Thanks to this, it's easier for me now*), we would say, agreeing with Sobol-Kwapińska and Oleś (2010), that in this case, we are dealing with the function of **redefining the past**.

In turn, in the dialogue between I in the present and I in the future, the latter seems to be fulfilling the role of the possible self, that is, the concept of self which presents a person in new social roles, in conditions that can happen, in a new environment, circumstances (Markus & Nurius, 1986). Anna's I in the future is the desired possible self (*I've got what I wanted. Professional success, I am fulfilling myself, I feel happy going to work and satisfied with the actions taken (…)*). The vision of the desired I in the future in the dialogue with a fairly uncertain I in the present (*Everything's not as easy as it looks from a different time perspective*) has primarily a **motivational function**. I in the future opens new perspectives to I in the present, mobilises it to action with a short statement (*but it's worth looking at the future and not worrying about what has been, just act*). Thanks to the supporting role of the imagined possible self in the future that spins plans and sets goals, we can talk about an attempt to **managing the future** as a function of this temporal dialogue.

However, in the context of the entire dialogue between the three temporal positions, the function of **building a distance** to what is "here and now" seems very important, which is clearly expressed by the metareflection (*In retrospect, everything seems simple. It's going to work*). Anna's dialogue finishes with a message expressing hope and optimism. It shows that detachment from the present allows for a different approach to life priorities (cf. Bandura, 2001).

Discussion

The aim of the paper was to answer questions about the main topics and functions of internal temporal dialogues in people during adolescence and early adulthood. The presented analyses have shown that in the temporal dialogues very different issues are raised, and at the same time, the topics are of great personal importance. We have also tentatively confirmed the theoretical proposal by Sobol-Kwapińska and Oleś (2010). The authors list eight functions of internal temporal dialogues. In our analysis, which included three examples of dialogues, we found seven of these functions. Of course, each of individual dialogues performed only selected functions. The first of the dialogues presented was conducive to balancing and redefining the past. The second provided support, advice and helped in acquiring wisdom. The third served to build a distance, manage the future and – similar to the first one – redefine the past. One function, that is making a decision, has not been illustrated in the analysed dialogues, although it cannot be excluded that such function is sometimes fulfilled by internal temporal dialogues. Moreover, it is possible that another functions are also performed by these dialogues. Thus, the question of functions of temporal dialogues needs further empirical and theoretical exploration.

In this context, before we discuss our results, we should emphasise the main limitations of our study. First, we used only qualitative analysis in order to explore the functions of temporal dialogues. Second, our analysis was based on the relatively small sample of dialogues. However, as it has been mentioned earlier the analysis presented here was the small part of a bigger project concerned different variables such as emotions and meaning of life. Because the other findings were presented elsewhere (Łysiak, 2017; Łysiak & Oleś, 2017), only the results of qualitative analysis could be shown here. In the future, it should be considered to prepare further qualitative as well as quantitative research on internal temporal dialogues, their determinants and functions. Additionally, the other age groups should be taken into account to maybe find out the individual differences on internal temporal dialogicality.

When discussing our results, it is worth noting that each of the presented dialogues ended with a metareflection – a summary which was a novum for the participant, and it was the result of juxtaposing, combining and interpreting voices coming from different temporal I-positions. It is also necessary to emphasise that not every participant of the study was able to generate a metareflection. In the light of the dialogical self theory, a person, being in a particular moment in the present, can try to look at themselves and the reality from the perspective of the past and the future. They can also trigger a dialogue of these temporal positions. But most importantly, as a result of this dialogue, they have a chance to look at their experiences from a broader perspective, as if from above, thus obtaining a whole new perspective on their present situation. In the dialogical self theory, this unique point of view is called metaposition or metareflection. Hermans and Hermans-Konopka (2010), compare the adoption of metaposition to the actions of a painter who, having finished their piece of art, steps back and studies the results with all the details, taking a broader perspective. This way, the artist builds a distance while still being in the moment of creating his artwork. According to the authors, metaposition has three basic functions: (1) provides a pervasive view of the multiplicity of positions; (2) enables a person to connect particular I-positions as elements of an overall story; (3) helps to find the direction of change. Hence the integration of voices as well as the unity of I-positions can occur (Hermans, 2001).

The analysis of internal temporal dialogues seems to confirm that a person performs metareflection usually after completing an important task in their life or closing one stage of life and symbolically opening the next one (Bandura, 2001; McAdams, 2010). A comprehensive look at past, present and future experiences can play a transgressive function – by learning new ideas and assimilating new experiences, an individual goes beyond the self-centred perspective (Kozielecki, 2007; Oleś et al., 2010). When writing about the adoption of the different temporal position Hermans (1996) suggests that with this capability we are able to perceive the meaning of the current actions. Making a metareflection by looking at life from a wider perspective contributes to summaries, balancing and revaluations. A metareflection allows a person to put all their experiences into one, connect them, it also allows to give meaning to their previous actions. Detaching from one point of view allows to change the way of thinking and sometimes also a course of action. The analysis of internal dialogues also shows that they become an opportunity to confront difficult, sometimes stressful issues, which consequently can be used to acquire wisdom and a mature view of both themselves and the world (cf. Oleś, 2011). The integration of temporal positions through metareflection can also be interpreted as an expression of the need to overcome a person's limitations, improve specific forms of activity and exceed their current capabilities. In this sense, it is an expression of the need for transcendence, generally understood as the ability to transcend a person's own conditions and expand the boundaries of the temporal self (cf. Worsch & Heckhausen, 2002). The metareflection resulting from the juxtaposition of voices of temporal positions is not just the integration of the temporal dimensions of the self. It also promotes the integration of the entire personality. A person has a need to understand past events, find a meaning in what he/she is doing and prepare for what is coming. As underlined by McAdams (1989, 1994), a human identity is a life story that integrates its reconstructed past, the perceived present and the anticipated future, giving it a sense of unity and a purpose of life. In this context, metareflection, which is a type of a link between contradictions and discrepancies in the interpretation of experiences seen from different temporal perspectives, allows the creation of identity.

Can the discussed temporal dialogues be connected with the increase of a feeling of happiness and well-being? Waytz, Hershfield, and Tamir (2015) in their research suggest that considering the past or the future allows people to transcend their day-to-day activities and to focus on the most important issues, which in turn is a potential source of the meaning of life. Also, Łysiak's study (2017; Łysiak & Oleś, 2017) shows that under the influence of temporal dialogue the meaning of life grows. In addition, the state of curiosity is intensifying, and in people who are able to finish their dialogue with a metareflection, the state of anxiety also decreases. These variables can be treated as selected measures of happiness. At the same time, they can be related to various pathways of pursuit of happiness. According to the hedonistic view (typical for e.g. Epicurean philosophy), pleasure is the main source of happiness. According to the eudaimonic approach view (propagated e.g. by Aristotelian philosophers) happiness is the result of engaging in valuable goals (Oleś & Jankowski, 2017). Currently, Seligman (2002; cf. Schueller & Seligman, 2010) in his authentic happiness theory combines hedonic and eudaimonic approaches. He posits three distinct pathways to well-being: apart from pleasure and engagement, he adds meaning. In this context, the reduction of anxiety after a temporal dialogue can be combined with pleasure, the increase of curiosity with engagement, while the increase of the sense of the meaning of life with the third pathway of pursuit of happiness, referred to as "meaning."

Considering the potential connection of temporal dialogues with well-being, it is also worth referring to studies on temporal orientation. Taking into account how people relate to time, Zimbardo and Boyd (1999) distinguished five types of temporal orientations: concentration on negative past (stress on traumas, disappointments, sad moments from the past); concentration on positive past (positive evaluation of the past); hedonistic concentration on present (stress on pleasure without considering the consequences); fatalistic concentration on present (the belief that attempts to influence the future are pointless), and concentration on future (formulating plans, setting goals). Several studies confirm that the types of time perspective are significantly linked to different important aspects of human functioning. For example, a past negative time perspective is connected with neuroticism, anxiety, depression, negative mood, low self-esteem, problems in social relations, gambling, and propensity for addiction (Klingemann, 2001; Stolarski, Matthews, Postek, Zimbardo, & Bitner, 2014; Zhang & Howell, 2011; Zimbardo & Boyd, 1999). Concentration on the fatalistic present correlates positively with risky behaviours, such as alcohol consumption and drug abuse (Daugherty & Brase, 2010; Keough, Zimbardo, & Boyd, 1999), whereas concentration on the future is connected with optimism (Zimbardo & Boyd, 1999) and health promoting behaviours (Boyd & Zimbardo, 2005).

Each of the presented internal temporal dialogues in its first part (I in the past vs. I in the present) reflects the concentration on the negative past, and also shows the elements of fatalistic treatment of the present, which may block the commitment to the future. However, the second part of the dialogue (I in the present vs. I in the future) favours the revaluation of the negative past or distancing from it and becoming involved in the future. In this context, a temporal dialogue can be treated as an attempt to achieve a so-called balanced time perspective, defined as a relatively strong concentration on the positive past, a moderate concentration on the future and the hedonistic present, as well as a weak concentration on the negative past and the fatalistic present (Zimbardo & Boyd, 1999). People with balanced time perspective are characterised by greater life satisfaction, less negative affect, and more frequent positive affect, a greater sense of the meaning in life, higher level of optimism, self-efficacy, happiness and mindfulness (Boniwell, Osin, Linley, & Ivanchenko, 2010; Drake, Duncan, Sutherland, Abernethy, & Henry, 2008). Thus, balanced time perspective is clearly connected with well-being.

Also, Shostrom (1974) emphasised the inadequacy of concentrating on only one of the three dimensions of time and used the term "time competence" as an essential element of a self-updating personality. In his opinion, self-updating people are able to link the past with the future in the present and they are less constrained by a sense of guilt, grief and anger from the past, while their aspirations entail realisation of the goals.

Can temporal internal dialogues that allow combining the three dimensions of time show connections with similar measures of well-being? The aforementioned studies by Łysiak (2017) and Waytz et al. (2015) suggest that indeed, but in order to give a precise answer to this question, further research is needed on internal temporal dialogues.

Disclosure statement

No potential conflict of interest was reported by the authors.

References

Assagioli, R. (2000). *Psychosynthesis: A collection of basic writings*. Amherst, MA: Synthesis Center.

Bandura, A. (2001). Social cognitive theory: An agentic perspective. *Annual Review of Psychology, 52*, 1–26.

Boniwell, I., Osin, E., Linley, P., & Ivanchenko, G. (2010). A question of balance: Examining relationships between time perspective and measures of well-being in the British and Russian samples. *Journal of Positive Psychology, 5*, 24–40.

Boyd, J. N., & Zimbardo, P. G. (2005). Time perspective, health and risk taking. In A. Strathman & J. Joireman (Eds.), *Understanding behavior in the context of time* (pp. 85–107). London: Lawrence Erlbaum.

Daugherty, J. R., & Brase, G. L. (2010). Taking time to be healthy: Predicting health behaviors with delay discounting and time perspective. *Personality and Individual Differences, 48*, 202–207.

Depape, A.-M. R., Hakim-Larson, J., Voelker, S., Page, S., & Jackson, D. L. (2006). Self-talk and emotional intelligence in university students. *Canadian Journal of Behavioural Science-Revue, 38*(3), 250–260.

Drake, L., Duncan, E., Sutherland, F., Abernethy, C., & Henry, C. (2008). Time perspective and correlates of well-being. *Time and Society, 17*, 47–61.

Hermans, H. J. M. (1996). Voicing the self: From information processing to dialogical interchange. *Psychological Bulletin, 119*, 31–50.

Hermans, H. J. M. (2001). The dialogical self: Toward a theory of personal and cultural positioning. *Culture Psychology, 7*(3), 243–281.

Hermans, H. J. M. (2003). The construction and reconstruction of a dialogical self. *Journal of Constructivist Psychology, 16*, 89–130.

Hermans, H. J. M., & Gieser, T. (Eds.). (2012). *Handbook of dialogical self theory*. Cambridge: Cambridge University Press.

Hermans, H. J. M., & Hermans-Jansen, E. (1995). *Self-narratives: The construction of meaning in psychotherapy*. New York: The Guilford Press.

Hermans, H. J. M., & Hermans-Konopka, A. (2010). *Dialogical self theory: Positioning and counter-positioning in a globalizing society*. Cambridge: Cambridge University Press.

Hermans, H. J. M., & Kempen, H. J. G. (1993). *The dialogical self: Meaning as movement*. San Diego: Academic Press.

Higgins, E. T. (1987). Self-discrepancy: A theory relating self and affect. *Psychological Review, 94*(3), 319–340.

Jacobsen, B. (2007). *Invitation to existential psychology: A psychology for the unique human being and its applications in therapy*. Chichester: John Wiley & Sons.

Keough, K. A., Zimbardo, P. G., & Boyd, J. N. (1999). Who's smoking, drinking, and using drugs? Time perspective as a predictor of substance use. *Journal of Basic and Applied Social Psychology, 21*, 149–164.

Klingemann, H. (2001). The time game: Temporal perspectives of patients and staff in alcohol and drug treatment. *Time & Society, 10*, 303–328.

Kozielecki, J. (2007). *Psychotransgresjonizm. Nowy kierunek w psychologii* [Psychotransgressionism: A new direction in psychology]. Warszawa: Wydawnictwo Akademickie „Żak".

Łysiak, M. (2017). Dialogical temporal chair technique. *British Journal of Guidance and Counselling*. doi:10.1080/03069885.2017.1413168

Łysiak, M., & Oleś, P. (2017). Temporal dialogical activity and identity formation during adolescence. *International Journal for Dialogical Science, 10*(1), 1–18.

Markus, H. R., & Nurius, P. (1986). Possible selves. *American Psychologist, 41*(9), 954–969.

McAdams, D. P. (1989). The development of a narrative identity. In D. M. Buss & N. Cantor (Eds.), *Personality psychology* (pp. 160–174). New York: Springer Verlag.

McAdams, D. P. (1994). *The person: An introduction to personality psychology*. Forth Worth: Hartcourt Brace College Publishers.

McAdams, D. P. (2010). The problem of meaning in personality psychology from the standpoints of dispositional traits, characteristic adaptations, and life stories. *The Japanese Journal of Personality, 18*(3), 173–186.

Neimeyer, R. A. (2006). Narrating the dialogical self: Toward an expanded toolbox for the counseling psychologist. *Counseling Psychology Quarterly, 19*(1), 105–120.

Oleś, P. K. (2011). *Psychologia człowieka dorosłego* [Psychology of an adult]. Warszawa: Wydawnictwo Naukowe PWN.

Oleś, P. K., Batory, A., Buszek, M., Chorąży, K., Dras, J., Jankowski, T., … Wróbel, M. (2010). Wewnętrzna aktywność dialogowa i jej psychologiczne korelaty [Internal dialogical activity and its psychological correlates]. *Czasopismo Psychologiczne, 16*, 113–127.

Oleś, P. K., & Jankowski, T. (2017). Positive orientation – a common base for hedonistic and eudemonistic happiness? *Applied Research in Quality of Life, 13*, 105–117.

Puchalska-Wasyl, M. (2015). Self-talk: Conversation with oneself? On the types of internal interlocutors. *The Journal of Psychology: Interdisciplinary and Applied, 149*(5), 443–460. doi:10.1080/00223980.2014.89677

Schueller, S. M., & Seligman, M. E. P. (2010). Pursuit of pleasure, engagement, and meaning: Relationships to subjective and objective measures of well-being. *Journal of Positive Psychology, 5*, 253–263.

Seligman, M. E. P. (2002). *Authentic happiness: Using the new positive psychology to realize your potential for lasting fulfillments*. New York: Free Press.

Shostrom, E. L. (1974). *Manual for the personal orientation inventory*. San Diego, CA: Educational and Industrial Testing Service.

Sobol-Kwapińska, M., & Oleś, P. K. (2010). Dialogi temporalne: Ja – w trzech wymiarach czasu [Temporal dialogues. Self – in three time dimensions]. In G. Sędek & S. Bedyńska (Eds.), *Życie na czas. Perspektywy badawcze spostrzegania czasu* [Life on time. Research perspectives on time perception] (pp. 398–420). Warszawa: Wydawnictwo Naukowe PWN.

Stolarski, M., Matthews, G., Postek, S., Zimbardo, P. G., & Bitner, J. (2014). How we feel is a matter of time: Relationships between time perspective and mood. *Journal of Happiness Studies, 15*, 809–827.

Waytz, A., Hershfield, H. E., & Tamir, D. I. (2015). Mental simulation and meaning in life. *Journal of Personality and Social Psychology, 108*(2), 336–355.

Worsch, C., & Heckhausen, J. (2002). Perceived control of life regrets: Good for young and bad for old adults. *Psychology and Aging, 17*, 340–350.

Zhang, J. W., & Howell, R. T. (2011). Do time perspectives predict unique variance in life satisfaction beyond personality traits? *Personality and Individual Differences, 50*, 1261–1266.

Zimbardo, P. G., & Boyd, J. N. (1999). Putting time in perspective: A valid reliable individual differences metric. *Journal of Personality and Social Psychology, 77*, 1271–1288.

Goal focused positive psychotherapy: an integration of positive psychology and psychotherapy

Evelyn I. Winter Plumb, Kathryn J. Hawley, Margaret P. Boyer, Michael J. Scheel and Collie W. Conoley 🔟

ABSTRACT

This article introduces the empirical support for and theoretical tenets of Goal Focused Positive Psychotherapy (GFPP), a comprehensive, evidence-based, psychotherapy model. GFPP's approach emerges from positive and social psychology research, and is informed by psychotherapy research from the common and contextual models. Its interventions focus on idiosyncratic and multiculturally-attuned client factors, particularly client strengths and goals, in an effort to increase subjective well-being and facilitate the client's experience of a meaningful, satisfying life. Enhanced client well-being provides more robust and abundant resources for proactively addressing presenting concerns without requiring intensive focus on client deficits, symptoms, or trauma. The mechanism of change is positive emotion, as informed by the broaden-and-build [Fredrickson, B. L. (2001). The role of positive emotions in positive psychology: The broaden-and-build theory of positive emotions. *American Psychologist*, *56*(3), 218–226. doi:10.1037/0003-066X.56.3.218] biopsychosocial model. The therapeutic alliance creates a healing context cultivated through a focus on hope, strengths, and client-centred experiences of self-determination.

GFPP's therapeutic mindset

A foundational distinction of Goal Focused Positive Psychotherapy (GFPP) is that GFPP therapists prioritise the promotion of well-being over the direct alleviation of symptoms, guided by the understanding that increasing psychological resources and active coping skills provides an equally-effective and approach-oriented pathway to managing presenting concerns. Approaches such as those found in traditional cognitive behavioural or psychodynamic treatment models orient the clinician to symptoms and histories of problems to provide clues for treatment. This deficit-oriented logic is grounded in a physical-sciences metaphor of change: a belief that identifying the symptom and its precipitants is the best way to reduce or eliminate the symptom. Consequently, treatment is often prescribed with a greater focus on the diagnosis and associated interventions than on the unique characteristics of a client (Hofmann & Hayes, 2018). In effect, a clinician practicing a deficit-oriented therapy narrows the focus of treatment to symptom distress and problem origin. Experiences of negative emotions such as anger or sadness are thoroughly investigated, presumably yielding a greater understanding of what contributes to client distress. While problem reduction can indeed occur through deficit-oriented approaches, achieving a happier, more meaningful life and flourishing is often a remote and inconsistent by-product of these forms of therapy.

In contrast, GFPP's mindset includes a commitment to expanding clients' perspectives in order to acknowledge and engage productively with a full range of negative and positive emotions, problems and exceptions to problems, and strengths and deficits. The GFPP clinician is keenly attentive to client experiences of positive emotions, desired states, and client strengths. Concentration is adjusted toward enhancing well-being, with the understanding that the "undoing" of the problem and its associated negative emotions will occur as a by-product of increased well-being (i.e. the undoing hypothesis of Broaden-and-Build; Fredrickson, Mancuso, Branigan, & Tugade, 2000).

Positive emotions are an important focus because they create experiences of broadening: they enhance cognitive flexibility, allowing clients to access a wider repertoire of solutions (Fredrickson, 2001). Instead of concentrating on avoidance of problem re-occurrence, clients are invited to contemplate and designate their desired states, which may occur separately from problems. The rationale for GFPP's approach is that by pursuing a happier, more meaningful life, clients increase their subjective well-being, and that this increase leaves the client better-equipped to effectively address the inevitable problems inherent to living.

Rather than the physical-sciences metaphor, GFPP incorporates a psychological metaphor for human change, based on the premise that people flourish under known conditions. These include a) involvement in meaningful activities; b) involvement in caring, supportive relationships; c) believing in one's efficacy; d) being focused on approach goals; and e) experiencing positive affect frequently (for a conceptual review, see Diener & Biswas-Diener, 2011). By promoting these five conditions of psychological change, clinicians can facilitate the cultivation of subjective well-being, fortifying clients to more effectively address life's challenges from a position of enriched psychological resources.

This tenet represents GFPP's use of psychological broadening as a curative factor, in contrast to a reliance on the narrowing effect of a concentration on problems and symptoms. Rather than a physical-sciences model in which the clinician delivers prescriptive interventions based on their expertise regarding the client's deficits and associated diagnoses, the psychological model engenders a non-hierarchical therapeutic stance of collaboration in which clinicians work alongside clients to help them create, invest in, and enjoy a more satisfying life by building on their strengths and contextual resources.

Clinician hope for their client is another crucial ingredient of GFPP's mindset. Clinician hope and client hope are linked (Coppock, Owen, Zagarskas, & Schmidt, 2010), and the clinician may experience hope before the client does that the client can lead a happier and more meaningful life. Importantly, clinician hope cannot be imitative; the clinician must authentically believe in the client's ability to achieve happiness. A growing body of scientific evidence highlights the significance of clinician's hope for their clients as a crucial condition for therapeutic change (Bartholomew et al., in press; Bartholomew, Gundel, & Scheel, 2017; Bartholomew, Scheel, & Cole, 2015; Scheel, Klentz Davis, & Henderson, 2013). Through a holistic concentration on client strengths, GFPP supports the clinician's authentic hope for their client, which in turn facilitates the client's hope for themselves and the therapeutic process.

GFPP's mechanism of change

Positive emotions have historically been overlooked by classic psychotherapy theories, as evidenced by the paucity of material on this topic in current psychotherapy textbooks (e.g. Corsini, 2001). Instead, psychotherapy has traditionally focused upon the cessation of psychological pain. An associated outcome of increased positive emotions may be implicitly assumed in many orthodox approaches to treatment, however it is rarely included as an outcome measure in traditional psychotherapy research. In contrast, GFPP unambiguously positions positive emotion as both an outcome goal and a principle process by virtue of its role as a mechanism of change.

The Broaden-and-Build theory (Fredrickson, 2001) elucidates the adaptive utility of positive emotions, and the associated research provides an empirical framework for their role as the facilitator of change in psychotherapy. The theory posits that positive emotions lead to psychological growth and well-being over time through two steps, broadening and building (Fredrickson, 2004). In the first step, when feeling positive emotions – such as joy, pride, interest, or love – individuals experience a broadened state of cognition that allows them to engage more flexibly with everyday life events. Although the neurological mechanisms underlying the effect of positive emotion on cognition remain unclear, a dopamaneric theory of positive affect suggests that positive emotions elicit a mild increase in the release of dopamine in the nigrostriatal system in the substantia nigra pars compacta (SN) and the mesocorticolimbic system in the ventral tegmental area (VTA; Ashby, Isen, & Turken, 1999). Dopamine projections from the SN and VTA to areas like the anterior cingulate and prefrontal cortex, among others, account for the subsequent observed increases in cognitive flexibility and creative problem solving (Ashby et al., 1999; Chiew & Braver, 2011).

In contrast to broadening, negative emotional experiences narrow thought-action repertoires to a specific set of action tendencies (Cosmides & Tooby, 2000). In times of danger and threat, the ability to be psychologically and physiologically primed for quick, specific action is clearly adaptive. In times of relative safety, however, perceiving more potential variables via enhanced cognitive flexibility allows individuals to view challenging situations in more engaged, novel, and creative ways (Fredrickson, 2004).

This enhanced potential for pursuing new experiences via heightened curiosity and flexibility (broadening), engenders the second step, building. When engaging with experiences in new and adaptive ways, individuals *build* their interpersonal and intrapersonal resources, yielding an expanded repertoire for future use (Fredrickson, 2004). These accumulated resources may be psychological, such as the ability to challenge negative automatic thoughts; physical, such as tactile ability or breath control; or interpersonal, such as new relationships or strengthened close bonds. Clients can use these accrued resources during times of stress and struggle, which promotes resilience and leads to improved long-term well-being (Cohn, Fredrickson, Brown, Mikels, & Conway, 2009).

To illustrate the change mechanism of positive emotions in therapy, consider a young female client struggling with social anxiety. She frequently feels discouraged by her difficulty connecting with others and interprets many everyday experiences as evidence that others are negatively evaluating her. Through her work with a GFPP clinician, she begins to recognise a passion for and strength in mathematics, which promotes feelings of pride and interest in session as well as during her math classes. In one such instance in class, the client found herself feeling pride for having received an A on her homework, then noticed that she had forgotten to bring a pencil for class. Buoyed by the swell of positive emotions from her recent success, she utilised her broadened problem-solving perspective by turning to the classmate behind her and asking to borrow a pencil – an experience that she had previously rated in session as "very" anxiety provoking. The classmate happily offered her a pencil, providing the client with new evidence that she is capable of asking for assistance from others and thereby widening her scope of plausible solutions for her next relevant challenge. The client enthusiastically described the experience to the clinician later that week as a successful interaction.

In the above example, the client was able to take advantage of a momentary opportunity for growth facilitated by positive emotion. In doing so, she gained new evidence that helps build her sense of confidence and perhaps encourages an approach–orientation towards future anxiety-provoking situations rather than falling back on her historically avoidant tendencies. Given that avoidance typically contributes to the long-term maintenance of anxiety, the alternative proactive strategy galvanised by this momentary experience of positive emotions could play a pivotal role in facilitating future successes and overall well-being.

The broaden-and-build theory has also been linked to an escalation effect, such that momentary experiences of positive emotion promote greater likelihood of future experiences of positive emotion. Because positive emotional experiences yield enduring psychological resources, individuals are able to engage with their environments with greater mastery, making it more likely that they will experience more frequent positive outcomes and subsequently, more positive emotions. These positive emotions may then in turn cultivate further resources that re-initiate the cycle, setting a positive escalation into motion (Fredrickson & Joiner, 2002).

Despite the broaden-and-build theory's prominence in the literature of experimental psychology, its direct applications to psychotherapy have been largely overlooked. Psychotherapy research has occasionally identified positive emotions as an *indicator* of change, but the research implications of broaden-and-build reach much further than outcomes: they provide an explanation for positive emotions' function as a primary facilitator of change in therapy (e.g. Fitzpatrick & Stalikas, 2008). This body of research posits that broadening effects constitute a common factor in psychotherapy, while the building effects constitute a common psychotherapeutic goal.

Positive emotions can be elicited during sessions in a variety of ways. The therapeutic relationship in particular provides an excellent medium for facilitating positive other-directed feelings, such as appreciation and caring. Clinicians may encourage more self-directed positive feelings such as interest and pride by inquiring about client hobbies, passions, and strengths, among other options. Conoley and Scheel (2018) provide a list of interventions to encourage positive emotional experiences in session. In these ways and more (as described below), Goal Focused Positive Psychotherapy represents a shift in the psychotherapy paradigm that elevates the role of positive emotions from an auxiliary and inconsistent by-product of treatment to an intentionally-cultivated mediator of client change.

Forming goals in GFPP

As the eponymous name suggests, client goals are the copestone of GFPP's interventional repertoire. Goals contain rich information about the client's implicit understanding of the Good Life and thus provide a kind of "meta-map" for treatment (Emmons, 2003). GFPP clinicians highlight the strengths that are revealed within client goals, and use them as implicit referents representing a client's internal vision of a life well-lived. A commitment to explicitly linking interventions and markers of progress to personally-meaningful goals ensures that the therapy process is continuously galvanised by client motivation while simultaneously grounding the therapeutic relationship in a mutual understanding of the client's desire for an experience of thriving defined in their own terms (Vansteenkiste, Simons, Lens, Sheldon, & Deci, 2004). Equally importantly, GFPP clinicians enthusiastically psycho-educate and support clients in setting goals that are comparatively more likely to reach fruition because they are characterised by the auspicious combination of intrinsically-motivating, self-determinative content and an approach orientation.

Intrinsic and extrinsic forces are often defined by their locus of impetus: Intrinsic having an internal origin and representing a valued end in itself, and Extrinsic having an external origin and representing more of an incentive to be attained than an action worthy of completion on its own merit (Hennessey, Moran, Altringer, & Amabile, 2005). Research on the motivational intensity and persistence of intrinsic versus extrinsic goal pursuit has compellingly demonstrated the superiority of the former (Ryan & Deci, 2000a), and this in itself is a pragmatic incentive to cultivate intrinsically-oriented goal content.

This benefit notably aligns with the three foundational psychological needs that most individuals require in order to thrive according to Self-Determination Theory (Ryan & Deci, 2000b): Competence, Autonomy, and Relatedness. The expression of these principles varies significantly depending on the cultural and temporal circumstances in which they arise, thus it is important to employ a contextually-sensitive perspective when mapping them onto client goals. Discovering the diverse ways in which these motifs emerge within the confluence of a client's socio-cultural and dispositional context can be a particularly exciting element of exploring a client's identity and social location as it relates to their goals. Highlighting and amplifying the themes of Autonomy, Competence, and Relatedness that are embedded within client goals not only fosters intrinsic motivation, but helps the client expand the foundational psychological elements necessary for them to thrive. Utilising Self-

Determination Theory as a referent for goal content additionally helps ensure that goals are freely-chosen, a critical component of the outcome-determining client factors of cooperation and motivation.

In addition to helping shape the content of the goal for optimal results, GFPP clinicians always aim to establish an approach-orientation towards the goal itself. An approach orientation describes a desire to create or enhance a positive outcome, whereas an avoidant orientation describes a desire to avoid, minimise, or ameliorate a negative outcome (Elliot & Friedman, 2007). A litany of research has demonstrated the advantages of approach goals over avoidant goals in terms of supporting motivation, morale, and persistence through challenges (e.g. Covington, 2000). A more nuanced yet also compelling benefit lies in the affective tone and quality of these outcomes: when a client accomplishes an avoidant goal, the most positive feeling they are likely to experience is relief – the absence of an anticipated negative emotion. Moreover, they are frequently burdened with the spectre of encountering the same challenge in the future, but without any accumulated evidence of their ability to effectively manage it other than simply continuing to evade it. Such an approach fails to reduce anxiety and often proliferates the symptoms by reinforcing the perception that the client's best strategy is to escape rather than encounter a challenge. Individuals who invest significant time and energy plotting how they can evade aversive experiences are likely to lead lives too narrowly circumscribed by their own fortress walls to allow in many opportunities for experiences of competence, mastery, or relatedness.

Rather than constructing elaborate psychological armour out of caveats and escape plans, an approach goal incrementally builds psychological resources by offering successive and open-ended opportunities for growth and mastery. Conspicuously paralleling Fredrickson's (2001) broaden and build spiral of Positive Emotions model, approach goals precipitate more creative thinking as well as more persistence, motivation, and flexibility (Jowkar, Kojuri, Kohoulat, & Hayat, 2014; Peng, Cherng, Chen, & Lin, 2013). These collateral benefits coalesce to form a goal-pursuit process that is not only more intrinsically enjoyable, but actually improves the likelihood of success: a true win-win scenario.

Client strengths in GFPP

GFPP embraces the concept that client factors are the overriding determinants of success in therapy, and this premise is most evident in the approach's emphasis on client strengths. Up to 86% of psychotherapy outcome is attributable to client life factors (Duncan, 2014), a term encompassing "all aspects idiosyncratic to the specific client and incidental to the treatment delivered" (Lambert, 2013; as quoted in Sparks & Duncan, 2016). Clearly, client strengths fall squarely within this purview, and GFPP clinicians view these resources as incomparable assets in the positive change process. This commitment to capitalising on the potency of client factors underscores GFPP's philosophical position that clients are far more active and powerful agents in their own recovery than has typically been acknowledged by diagnostically-focused and deficit-based psychotherapy models (Bohart & Wade, 2013).

While strengths have been scrupulously studied and operationalised in varied forms in the positive psychology and psychotherapy literature, in GFPP the term is used voluminously to refer to "anything that a client has done, believed, or felt that can facilitate the client's growth in a healthy way" (Conoley & Scheel, 2018). This flexible definition allows the client and the clinician to collaborate in determining what has already served the client in their growth and well-being, as well as how those assets may be implemented in novel and creative ways to overcome challenges and further augment the client's life.

Client-centred exploration within a collaborative framework is absolutely critical in the identification and mobilisation of strengths, as this domain is notoriously susceptible to clinician misinterpretation, particularly within cross-cultural exchanges (Wong, 2006). It is essential for clinicians to elicit, acknowledge, and privilege the client's understanding of what does and does not constitute a positive character trait or resource for their own individual circumstances, in order to avoid impos- ing clinicians' moral and cultural belief systems onto their client. For example, a trait such as confi- dence or extroversion may be lauded as a strength in one culture but considered a character flaw in another, and making assumptions about how the client appraises a particular trait can result in significant decrements to the therapeutic alliance. In contrast, respectfully sorting through the intri- cacies of context and culture to determine the valence of a client's particular trait can stimulate a dynamic exploratory dialogue between client and clinician and contribute to a richer intersubjective experience of the therapeutic relationship.

While always foregrounding the client's cultural-autobiographical context in highlighting strengths, a GFPP clinician may also use classic psychotherapeutic interventions such as Reframing, Success-Finding (GFPP's version of Solution-Focused Brief Therapy's "Exception Finding" technique, Conoley & Scheel, 2018), Encouragement, Open-Ended questions about cultural and familial heroes/role models, and many other techniques in order to reveal previously-unrecognized client strengths. When care is taken to ensure that the identified strengths are personally and culturally-syn- tonic, these interventions serve a central GFPP goal of generating a more positive client self-image and an expanded, more accessible repertoire of intrapersonal resources. For example, it is not uncom- mon for client strengths to emerge in the context of adversity, and drawing attention to the resources that the client is using in order to confront and move past a challenge can help them to form a more integrated, balanced, and empowering understanding of themselves and their autobiographical nar- rative of a difficult event than one in which they solely acknowledge the obstacles.

Consider a client who presents for therapy feeling deeply demoralised and concerned about her financial and vocational future after being laid off from several jobs in a row. While acknowledging and attentively validating the significant distress and anxiety associated with these events, the GFPP clinician also emphatically highlights: the resourcefulness that the client has utilised in making ends meet for her family on a shoestring budget, the courage that she has shown by persisting in the face of discouraging circumstances, and the love for her children and commitment to her community that she has displayed by using her newfound extra time to volunteer at school and advocate locally for workers' rights.

This process of spotlighting strengths opens a pathway to exploring the client's intersecting iden- tities and corresponding cultural and personal strengths. She experiences the positive emotions of pride and hope while sharing about how she learned techniques for creative budgeting from her maternal grandmother, who grew up during the Great Depression, and how she was instilled with the value of civic pride and the importance of social justice from her father, an agricultural labourer who protested with César Chávez for workers' rights when she was a young child. Without minimising the scope of her challenge, this new resource-centred perspective shifts the client's view of her situ- ation to a more empowering one in which her strengths, autonomy, values, competence, and intrinsic motivation are foregrounded, rather than one in which she views herself primarily as a passive casualty of unfortunate circumstances. The brief *in situ* experience of positive emotions that she experiences in reflecting on these strengths spurs a broadening and building process in which she begins to explore how she might follow her father's example of going to college to make a later- life career change, a possibility which previously felt too overwhelming for her to consider.

Such interventions and their corollary outcomes demonstrate GFPP's position that clients are best- equipped to enjoy life and respond effectively to challenges when they are aware and appreciative of their own strengths. Personal strengths may not fully obviate distress or difficulties on their own, but they certainly create a more empowered and abundant context of personal resources from which clients may engage with inevitable life obstacles. Notably, the benefits of highlighting client strengths appear to be a synergistic process: Flückiger and Grosse Holtforth (2008) found that client outcomes were significantly enhanced when clinicians were reminded of client strengths prior to each session, and Conoley and Scheel (2018) note that clinician hope for their clients – as instilled by a focus on client strength – can serve as the tinder that ignites client hope for themselves.

Hope in GFPP

In the GFPP model, hope serves a central purpose in the psychotherapeutic process, and a GFPP clinician endeavours to enhance client hope through several avenues. As a powerful positive emotion, hope galvanises change through broadening and building upon client resources (Fredrickson, 2001), enhancing belief in the change process (Frank & Frank, 1993), and boosting well-being (Alarcon, Bowling, & Khazon, 2013).

Hope is widely regarded as a significant common factor in psychotherapy. Frank and Frank (1993) suggested that a fundamental purpose of psychotherapy is to provide hope for clients, who typically arrive to treatment feeling demoralised. In their Contextual Model, Frank and Frank proposed that client re-moralization occurs through a collaborative therapeutic relationship and a compelling rationale that enhances positive client outcome expectancies. In GFPP, hope is initially cultivated by providing clients with a transparent rationale for the emphasis on positive processes. The reasoning behind the approach is discussed explicitly so that clients feel involved in a collaborative, respectful, and empowering process. The clinician draws on the client's unique parlance and perspective to provide justification for interventions, aiming to increase client belief in the change process, and thus enhance hope for positive outcomes.

The GFPP relationship itself also serves as a facilitator of hope. Client hope for change through counselling is positively related to the therapeutic alliance (Bartholomew et al., 2015), and client hope is enhanced when clinicians model hopeful thoughts and behaviours (Snyder, 1994). GFPP helps clinicians maintain hope for their clients through the focus on client strengths and potential rather than deficits and barriers.

Snyder's Hope Theory asserts that hope is composed of three crucial elements: goals, pathways thinking, and agency. When applied to psychotherapy, Hope Theory suggests that clients must have a target for therapeutic change (i.e. a goal), a plan for how that change will occur (i.e. pathways thinking), and a sense that they are competent and capable of affecting change in their lives (i.e. agency) (Snyder, 2002). GFPP clinicians purposefully highlight all three of these elements and bring them explicitly into client awareness.

GFPP treatment begins with exploring and identifying approach goals and unearthing intrinsic motivations for goal selection. This process facilitates the designation of an important target for therapeutic change, and reminds the client of what makes that target meaningful and motivating. To enhance pathways thinking, clients are encouraged to draw from their own wisdom regarding how to best facilitate movement toward goals, often via success-finding and solution-focused interventions. Pathways thinking is expanded by GFPP's focus on the psychological model, and the corresponding belief that a desired outcome can be reached via multiple psychological "routes". Finally, client agency is supported through reflection of client strengths, and celebration of small successes and incremental steps toward change. Recalling successes enhances client agency and motivation to work toward further change (Cheavens, Feldman, Woodward, & Snyder, 2006).

Intrinsic motivation is necessary in order for clients to continue progressing toward goals in psychotherapy (Ryan, Lynch, Vansteenkiste, & Deci, 2011). GFPP incorporates Self Determination Theory and its principles of competence, relatedness, and autonomy to address the determinants of motivation and their relationship to hope in psychotherapy. In GFPP, client competence – like agency – is emphasised through reflection of strengths and the ostensive celebration of incremental steps toward goals. Relatedness is effectuated through a strong, warm, collaborative therapeutic relationship. Autonomy is encouraged through the clinician's support of goals that feel personally relevant and culturally-syntonic to the client, as described above.

This summary outlines just a few of the most prominent access points for cultivating and capitalising on client hope. Many additional interventional avenues are available for clinicians aiming to enhance their clients' sense of hope and draw on all the corollary benefits; exploring this process with the client can be an informative, dynamic, and creative component of treatment.

Research supporting GFPP

The effectiveness of GFPP is supported by research on common factors, a three-year effectiveness study, case studies, and a qualitative study, as described below. Common factors research supports GFPP through a number of meta-analyses supporting the argument that psychotherapy approaches that include the common factors attain equivalent outcomes, and the common factors are necessary and sufficient for change (e.g. Wampold & Imel, 2015). GFPP includes these factors (Conoley & Scheel, 2018), which are: (a) a therapeutic emotional bond, (b) a therapeutic setting in which client self-disclosure occurs, (c) a psychologically and culturally acceptable rationale for change that addresses the client's issues, and (d) procedures that help the client (Frank & Frank, 1993; Wampold & Imel, 2015). Lambert (2013) further asserts that therapy models containing the common factors will be efficacious for the psychological presenting problem.

Direct support for the effectiveness of GFPP was established by comparing process and outcome measures for GFPP with those of Cognitive Behavioural Therapy (CBT) and Psychodynamic-Interpersonal therapy treatments in a field study (Conoley & Scheel, 2018). All three treatment models involved expert supervisors and doctoral clinicians-in-training dedicated to performing the respective treatments. The sample included 91 adult clients with a mean age of 27. Sixty-six percent of the clients self-identified as female and 44% self-identified as male, with 54% identifying as European American, 16% Latinx, 9% Asian American, 11% "Other", and 10% declining to identify their ethnicity. The training clinic accepted clients with all diagnoses except active psychosis or drug/alcohol dependence.

As expected from the common factors research, no significant outcome differences were found between the treatment groups using the Outcome Questionnaire-45 (Lambert et al., 1996) and the Outcome Rating Scale (Duncan & Reese, 2015). However, GFPP was rated significantly higher in the domain of therapeutic alliance (Session Rating Scale, Duncan et al., 2003). Compared to the CBT and Psychodynamic-Interpersonal treatment approaches, the clients rated GFPP higher in: (a) feeling heard, understood, and respected; (b) working on and talking about things that were important to the client; (c) believing that the approach personally fit; and (d) generally thinking that the session was good for the client. These results were consistent with GFPP theory's prediction that the focus on client strengths and approach goals would render the experience of therapy more appealing and the relationship with the clinician more positive in comparison to traditional approaches.

The effectiveness of GFPP is also supported by multiple case studies. One case study accompanies this article (reference redacted). Four case studies are included in "Goal Focused Positive Psychotherapy: A strengths based approach" (Winter Plumb, Byrom, Bartholomew, & Hawley, 2018). Case studies perform several important functions in the research literature (Yin, 1994). This type of research utilises actual client-clinician interactions to exemplify the functioning of complex processes in context. The examples extend GFPP descriptions into dyadic exchanges that deepen understanding and decrease misconceptions by vivifying the technical descriptions in theory and quantitative research articles with real-life, *in situ* examples. Furthermore, case studies demonstrate that the processes asserted to occur in a psychotherapy model do in fact occur. Because the clinician and client create the healing factors within their dyadic context, the case study reader receives a richer understanding of the possible applications by integrating their technical understanding of the approach with the qualitative substrate provided by case studies. Case studies are especially relevant for this approach because GFPP focuses upon client factors; the idiosyncratic strengths of each client which cannot be adequately represented in larger-scale quantitative formats. Each of the case studies provides validity to the claims of GFPP in terms of outcome and process.

Finally, a qualitative study (Hawley & Winter Plumb, 2016) revealed that clients overwhelmingly experienced GFPP as facilitating personal growth and goal attainment, as well as mastering new skills and perspectives. Clients recognised the processes attributed to GFPP, and stated that negative emotions and experiences were addressed thoroughly and effectively alongside the focus on positive emotions and experiences.

In summary, the effectiveness of GFPP is supported by direct and indirect evidence. The qualitative study, case studies, and outcome evaluation of GFPP directly support the effectiveness, while meta-analyses supporting the effectiveness of psychotherapies containing common factors provide indirect support.

Disclosure statement

No potential conflict of interest was reported by the authors.

ORCID

Collie W. Conoley ⓘ http://orcid.org/0000-0003-1793-8719

References

Alarcon, G. M., Bowling, N. A., & Khazon, S. (2013). Great expectations: A meta-analytic examination of optimism and hope. *Personality and Individual Differences, 54*(7), 821–827. doi:10.1016/j.paid.2012.12.004

Ashby, F. G., Isen, A. M., & Turken, A. U. (1999). A neuropsychological theory of positive affect and its influence on cognition. *Psychological Review, 106*(3), 529–550. doi:10.1037/0033-295X.106.3.529

Bartholomew, T. T., Gundel, B. E., Li, H., Joy, E. E., Kang, E., & Scheel, M. J. (under review). 'They don't have hope … they're looking to us': The phenomenology of therapists' hope for clients.

Bartholomew, T. T., Gundel, B. E., & Scheel, M. J. (2017). The relationship between alliance ruptures and hope for change through counseling: A mixed methods study. *Counselling Psychology Quarterly, 30*, 1–19. doi:10.1080/09515070.2015.1125853

Bartholomew, T. T., Scheel, M. J., & Cole, B. P. (2015). Development and validation of the hope for change through counselling scale. *The Counselling Psychologist, 43*(5), 671–702. doi:10.1177/0011000015589190

Bohart, A., & Wade, A. G. (2013). The client in psychotherapy. In M. J. Lambert (Ed.), *Bergin and Garfield's handbook of psychotherapy and behaviour change* (6th ed., pp. 219–257). Hoboken, NJ: Wiley.

Cheavens, J. S., Feldman, D. B., Woodward, J. T., & Snyder, C. R. (2006). Hope in cognitive psychotherapies: On working with client strengths. *Journal of Cognitive Psychotherapy, 20*, 135–145. doi:10.1891/088983906780639844

Chiew, K. S., & Braver, T. S. (2011). Positive affect versus reward: Emotional and motivational influences on cognitive control. *Frontiers in Psychology, 2*, 279. doi:10.3389/fpsyg.2011.00279

Cohn, M. A., Fredrickson, B. L., Brown, S. L., Mikels, J. A., & Conway, A. M. (2009). Happiness unpacked: Positive emotions increase life satisfaction by building resilience. *Emotion, 9*(3), 361–368. doi:10.1037/a0015952

Conoley, C. W., & Scheel, M. J. (2018). *Goal focused positive psychotherapy: A strengths-based approach.* NY: Oxford University Press. doi:10.1093/med-psych/9780190681722.001.0001

Coppock, T. E., Owen, J. J., Zagarskas, E., & Schmidt, M. (2010). The relationship between therapist and client hope with therapy outcomes. *Psychotherapy Research, 20*, 619–626. doi:10.1080/10503307.2010.4979508

Corsini, R. J. (2001). *Handbook of innovative therapy.* Hoboken, NJ: John Wiley & Sons.

Cosmides, L., & Tooby, J. (2000). Evolutionary psychology and the emotions. In M. Lewis & J. M. H. Jones (Eds.), *Handbook of emotions* (pp. 91–115). New York: Guilford.

Covington, M. V. (2000). Goal theory, motivation, and school achievement: An integrative review. *Annual Review of Psychology, 51*(1), 171–200. doi:10.1146/annurev.psych.51.1.171

Diener, E., & Biswas-Diener, R. (2011). *Happiness: Unlocking the mysteries of psychological wealth.* Malden, MA: John Wiley & Sons.

Duncan, B. L. (2014). *On becoming a better therapist: Evidence-based practice One client at a time* (2nd ed.). Washington, DC: American Psychological Association.

Duncan, B. L., Miller, S. D., Sparks, J. A., Claud, D. A., Reynolds, L. R., Brown, J., & Johnson, L. D. (2003). The session rating scale: Preliminary psychometric properties of a 'working' alliance measure. *Journal of Brief Therapy, 3*(1), 3–12.

Duncan, B. L., & Reese, R. J. (2015). The partners for change outcome management system (PCOMS): Revisiting the client's frame of reference. *Psychotherapy, 52*(4), 391–401. doi:10.1037/pst0000026

Elliot, A. J., & Friedman, R. (2007). Approach and avoidance personal goals. In B. Little, K. Salmela-Aro, J. Nurmi, & S. Phillips (Eds.), *Personal project pursuit: Goals, action, and human flourishing* (pp. 95–118). Hillsdale, NJ: Lawrence Erlbaum Associates.

Emmons, R. A. (2003). Personal goals, life meaning, and virtue: Wellsprings of a positive life. In C. L. M. Keyes & J. haidt (Eds.), *Flourishing: positive psychology and the life well-lived* (pp. 105–128). Washington, DC: American Psychological Association.

Fitzpatrick, M. R., & Stalikas, A. (2008). Positive emotions as generators of therapeutic change. *Journal of Psychotherapy Integration, 18*(2), 137–154. doi:10.1037/1053-0479.18.2.137

Flückiger, C., & Grosse Holtforth, M. (2008). Focusing the therapist's attention on the patient's strengths: A preliminary study to foster a mechanism of change in outpatient psychotherapy. *Journal of Clinical Psychology, 64*(7), 876–890. doi:10.1002/jclp.20493

Frank, J. D., & Frank, J. B. (1993). *Persuasion and healing: A comparative study of psychotherapy* (3rd ed.). Baltimore, MD: Johns Hopkins University Press.

Fredrickson, B. L. (2001). The role of positive emotions in positive psychology: The broaden-and-build theory of positive emotions. *American Psychologist, 56*(3), 218–226. doi:10.1037/0003-066X.56.3.218

Fredrickson, B. L. (2004). The broaden-and-build theory of positive emotions. *Philosophical Transactions of the Royal Society B: Biological Sciences, 359*(1449), 1367–1377. doi:10.1098/rstb.2004.1512

Fredrickson, B. L., & Joiner, T. (2002). Positive emotions trigger upward spirals towards emotional well-being. *Psychological Science, 13*(2), 172–175. doi:10.1111/1467-9280.00431

Fredrickson, B. L., Mancuso, R. A., Branigan, C., & Tugade, M. M. (2000). The undoing effect of positive emotions. *Motivation and Emotion, 24*(4), 237–258. doi:10.1023/A:1010796329158

Hawley, K. J., & Winter Plumb, E. I. (2016, August). Client responses to goal-focused positive psychotherapy. In B. Cole (Chair), *Goal focused positive psychotherapy: A comprehensive positive psychology therapy model.* Symposium conducted at the annual convention of the American Psychological Association, Denver, CO.

Hennessey, B., Moran, S., Altringer, B., & Amabile, T. M. (2005). Extrinsic and intrinsic motivation. *Wiley Encyclopaedia of Management.* doi:10.1002/9781118785317.weom110098

Hofmann, S. G., & Hayes, S. C. (2018). The future of intervention science: Process-based therapy. *Clinical Psychological Science.* doi:2167702618772296

Jowkar, B., Kojuri, J., Kohoulat, N., & Hayat, A. (2014). Academic resilience in education: The role of achievement goal orientations. *Journal of Advances in Medical Education & Professionalism, 2*(1), 33–38. Retrieved from www.ncbi.nlm.nih.gov/pmc/articles/PMC4235534/

Lambert, M. J. (2013). The efficacy and effectiveness of psychotherapy. In M. J. Lambert (Ed.), *Bergin and Garfield's handbook of psychotherapy and behaviour change* (6th ed., pp. 169–218). New York: Wiley.

Lambert, M. J., Hansen, N. B., Umphress, V., Lunnen, K., Okiishi, J., Burlingame, G. M., & Reisinger, C. W. (1996). *Administration and scoring manual for the outcome questionnaire (OQ45.2).* Wilmington, DE: American Professional Credentialing Services.

Peng, S. L., Cherng, B. L., Chen, H. C., & Lin, Y. Y. (2013). A model of contextual and personal motivations in creativity: How do the classroom goal structures influence creativity via self-determination motivations? *Thinking Skills and Creativity, 10*, 50–67. doi:10.1016/j.tsc.2013.06.004

Ryan, R. M., & Deci, E. L. (2000a). Intrinsic and extrinsic motivations: Classic definitions and new directions. *Contemporary Educational Psychology, 25*(1), 54–67. doi:10.1006/ceps.1999.1020

Ryan, R. M., & Deci, E. L. (2000b). Self-determination theory and the facilitation of intrinsic motivation, social development, and well-being. *American Psychologist, 55*(1), 68–78. doi:10.1037/0003-066X.55.1.68

Ryan, R. M., Lynch, M. F., Vansteenkiste, M., & Deci, E. L. (2011). Motivation and autonomy in counselling, psychotherapy, and behaviour change: A look at theory and practice. *The Counselling Psychologist, 39*(2), 193–260. doi:10.1177/0011000009359313

Scheel, M. J., Klentz Davis, C., & Henderson, J. (2013). Therapist use of client strengths: A qualitative investigation of positive processes. *The Counselling Psychologist, 41*(3), 392–427. doi:10.1177/0011000012439427

Snyder, C. R. (1994). *The psychology of hope: You can get there from here.* New York: Free Press. doi:10.1207/S15327965PL1304_01

Snyder, C. R. (2002). Hope theory: Rainbows in the mind. *Psychological Inquiry, 13*(4), 249–275. doi:10.1207/S15327965PLI1304_01

Sparks, J. A., & Duncan, B. L. (2016). Client strengths and resources: Helping clients draw on what they already do best. In M. Cooper & W. Dryden (Eds.), *The handbook of pluralistic counselling and psychotherapy* (pp. 68–79). London: Sage.

Vansteenkiste, M., Simons, J., Lens, W., Sheldon, K. M., & Deci, E. L. (2004). Motivating learning, performance, and persistence: The synergistic effects of intrinsic goal contents and autonomy-supportive contexts. *Journal of Personality and Social Psychology, 87*(2), 246–260. doi:10.1037/0022-3514.87.2.246

Wampold, B. E., & Imel, Z. E. (2015). *The great psychotherapy debate: The evidence for what makes psychotherapy work.* Routledge. doi:10.4324/9780203582015

Winter Plumb, E. I., Byrom, R., Bartholomew, T. T., & Hawley, K. (2018). Goal focused positive psychotherapy case examples. In C. W. Conoley & M. J. Scheel (Eds.), *Goal focused positive psychotherapy: A strengths-based approach* (pp. 97–122). NY: Oxford University Press. doi:10.1093/med-psych/9780190681722.001.0001

Wong, Y. J. (2006). Strength-centred therapy: A social constructionist, virtues-based psychotherapy. *Psychotherapy: Theory, Research, Practice, Training, 43*(2), 133–146. doi:10.1037/0033-3204.43.2.133

Yin, R. K. (1994). *Case study research: Design and methods* (2nd ed.). Thousand Oaks, CA: Sage.

The effect of positive psychology interventions on hope and well-being of adolescents living in a child and youth care centre

Krysia Teodorczuk, Tharina Guse ⓘ and Graham A du Plessis

ABSTRACT
This study evaluated the effect of positive psychology interventions (PPIs) on hope and well-being among adolescents living in a child and youth care centre (CYCC) in South Africa. Adolescents ($n = 29$) were allocated to either the experimental or control group through matched sampling. The experimental group engaged in one-hour intervention sessions weekly for six weeks. Measures of well-being and hope were recorded at three time intervals. Independent- and paired-sample t-tests were conducted to establish group differences. There were no statistically significant differences in well-being and hope between the two groups after the interventions. We discuss moderating factors and offer a qualitative reflection to better understand these outcomes. With this understanding, preliminary guidelines are proposed for implementing PPIs in CYCCs.

Adolescence is a transitional period of life involving rapid physical, biological, cognitive and psychosocial development (Berger, 2011; Santrock, 2010). These substantial changes contribute towards a notoriously tumultuous and challenging decade of life (Santrock, 2010). Adolescents living in Child and Youth Care Centres (CYCCs) experience stressors beyond those of their family-nurtured peers. Because of a history of complex and frequently maladaptive home, school and social environments, these youths are vulnerable to increased developmental challenges and psychological malady.

Research has repeatedly revealed significantly higher levels of psychopathology (Kjelsberg & Nygren, 2004; Richardson & Lelliott, 2003) and lower levels of well-being (Leslie, Gordon, Ganger, & Gist, 2002; Zimmer & Panko, 2006) among residents of youth care facilities when compared to their family-nurtured peers. Moreover, these vulnerable youths are also prone to low levels of self-worth, self-esteem and hope towards positive outcomes for their future (Aguilar-Vafaie, Roshani, Hassanabadi, Masoudian, & Afruz, 2011; Milkman & Wanberg, 2012). In combination, increased psychopathology, reduced well-being and low levels of hope reduce vulnerable youths' prognosis for optimal functioning.

Traditionally, research on mental health of youths in general and more specifically vulnerable youths focussed almost exclusively on psychological disorder (Evans et al., 2005). However, interventions associated with this framework gave no attention to individuals' positive attributes or strengths (Seligman, 2002; Seligman & Csikszentmihalyi, 2000). Yet, it is equally important to promote mental health and to encourage optimal functioning through a focus on that which is positive, adaptive and whole (Peterson & Seligman, 2004). Interventions grounded in positive psychology offer a manner in which to support and enhance positive growth through identification, utilisation and cultivation of psychological strengths, and thus present a means to enable and empower young people.

Further, due to the maturation of cognitive capacities, adolescence offers a timeous opportunity to introduce tools and exercises that may promote growth, cultivate strengths and enhance well-being even in the presence of vulnerability. It is therefore valuable to implement and evaluate positive interventions among youth living in CYCCs.

The social contexts within which adolescents grow up play an important role in their development. These environments and the associated circumstances contribute towards protective and risk factors that promote or hinder the transition to adulthood. In particular, poverty, long-term family disadvantage, domestic violence and abuse, maltreatment, neglect and abandonment, as well as multiple placements in out of home facilities, and lack of family contact are among established risk factors known to predict negative outcomes for developing adolescents (Berger, 2011; Cluver & Gardner, 2007; Coleman & Hagell, 2007).

The high occurrence of individual, family and community risk factors that impact on looked-after adolescents, place them at high risk for a poor prognosis involving social, psychological and behavioural problems culminating in restricted life opportunities (Coleman & Hagell, 2007; Fergusson & Horwood, 2003). However, some individuals display remarkable resilience and despite their exposure to major adversity, appear to cope well and adapt positively. These resilient youths are able to draw from various resources that may assist them to resist significant stress, trauma and adversity. The reservoir of resources, including individual, family and community attributes are referred to as protective factors (Aguilar-Vafaie et al., 2011; Coleman & Hagell, 2007).

Research focussed on optimal youth development draws the distinction between protective factors at an individual level as well as those at a social-contextual level that include family and community attributes (Aguilar-Vafaie et al., 2011; Coleman & Hagell, 2007). Cluver and Gardner (2007) indicated that little research has been done regarding protective factors among looked-after adolescents in the South African context. However, they reported that familial and social protective factors including pro-social peer relationships, and positive activities including sport, dancing and reading were associated with lower levels of anxiety and depression. International research has identified individual protective factors in at-risk children including attributes such as hope (Hagen, Myers, & Mackintosh, 2005; Herth, 1998), positive views of the self (Cicchetti, 2010; Coleman & Hagell, 2007; Daniel, Wassell, & Gilligan, 1999; Fergusson & Horwood, 2003), goal-setting and achievement orientation (Cicchetti, 2010; Coleman & Hagell, 2007), pro-social values, relationships and behaviours (Aguilar-Vafaie et al., 2011; Cicchetti, 2010; Daniel et al., 1999; Milkman & Wanberg, 2012; Mullan & Fitzsimons, 2006), the development of skills (Coleman & Hagell, 2007; Mullan & Fitzsimons, 2006), as well as identification and use of character strengths (Moore, 2010; Park, 2004). Enhancing these attributes may serve to buffer vulnerable youth against stressful and negative life circumstances.

One way to enhance such positive attributes is through implementing positive psychology interventions (PPIs). PPIs are intentional activities aimed to cultivate positive emotions, cognitions or behaviours (Sin & Lyubomirsky, 2009). These interventions are usually brief and easy to implement, yet are theory-informed and evidence based (Parks & Biswas-Diener, 2013). In this study, we selected PPIs that aimed to identify and build character strengths, and to cultivate positive emotions and hope.

Character strength interventions are based on the premise that developing individuals' strengths rather than focussing on their weaknesses may produce greater benefits (Quinlan, Swain, & Vella-Brodrick, 2012). Character strengths are positive traits reflected in an individual's thoughts, feelings and behaviours (Park, Peterson, & Seligman, 2004). Engaging one's character strengths allows individuals to achieve optimal functioning while pursuing valued endeavours. As such, character strengths act as protective factors that buffer youths against adverse life circumstances, and also support and enhance well-being (Quinlan et al., 2012).

The maladaptive home and social environments from which adolescents residing in CYCCs hail, often lack support and guidance, inhibiting them from identifying and developing character strengths (Epstein, 2000). Research has shown that identifying, planning to develop, using and building character strengths may increase well-being and alleviate depression (Gander, Proyer, Ruch, & Wyss, 2012; Proctor et al., 2011; Seligman, Steen, Park, & Peterson, 2005). Therefore, it is important to give these youths a general understanding of character strengths, and provide them with the opportunity to identify, utilise and cultivate their strengths. Strengths-based interventions include "you at your best" (Seligman et al., 2005), "strength spotting" (Proctor et al., 2011) and "using character strengths in a new way" (Seligman et al., 2005).

Activities that *generate positive emotions* give rise to an upward spiral of positive emotions, open up new possibilities, potentially broaden behaviour and build psychological resources (Fredrickson, 2001). Gratitude activities can be good starter exercises to facilitate the experience of positive emotions (Layous, Lee, Choi, & Lyubomirsky, 2013). Research indicated that gratitude interventions enhance well-being, life satisfaction, and optimism as well as reduce negative affect and lower levels of depression and anxiety (Emmons & McCullough, 2003; Froh, Sefick, & Emmons, 2008; Seligman et al., 2005).

Engaging in positive aspects of the present moment can also induce an upward spiral of emotions that ultimately enriches life through enhancing well-being while presenting an opportunity to thrive (Guse, 2014). Doing acts of kindness, savouring and loving-kindness meditation (LKM) are activities that focus on enhancing positive emotions in the present moment. Evidence suggests that pro-social activity, including acts of kindness, may buffer vulnerable adolescents from engaging in disruptive and problem behaviours (Milkman & Wanberg, 2012). Further, such altruistic behaviour may serve to build resilience, self-respect and hopefulness, which may protect vulnerable youths against feelings of defeat and despair. Additionally, these well-intended behaviours can enhance social relationships and increase well-being (Guse, 2014; Suldo & Michalowski, 2007).

Savouring is a process through which people "attend to, appreciate, and enhance the positive experiences in their lives" (Bryant & Veroff, 2007, p. 2). Immersion in favourable experiences strengthens awareness of the positive and pleasurable, which simultaneously reduces negativity bias and defends against negative emotions (Suldo & Michalowski, 2007). Intervention studies implementing savouring exercises reported positive correlations with well-being (Bryant, Smart, & King, 2005; Jose, Lim, & Bryant, 2012) self-esteem, positive affect and life satisfaction (Cafasso, 1994, 1998), as well as negative correlations with negative affect and depressive symptoms (Cafasso, 1994, 1998; Hurley & Kwon, 2012).

Like savouring, LKM also offers an opportunity for enhancing positive emotions in the present moment. This meditative practice has its origins in Buddhist traditions of emphasising and cultivating connectedness, whilst expressing positive and loving intentions towards others (Hutcherson, Seppala, & Gross, 2008). Studies that implemented LKM in the context of social and interpersonal relationships demonstrated enhanced positive affect, compassion and empathy, promoted optimism and reduced negative affect (Fredrickson, Cohn, Coffey, Pek, & Finkel, 2008; Garland et al., 2010; Hutcherson et al., 2008; Kristeller & Johnson, 2005).

Hope-based interventions focus on generating positive future expectations. Increasing levels of hope through specific activities may enhance well-being, positive affect and goal-directed activity whilst reducing symptoms of depression and anxiety (Cheavens, Feldman, Gum, Michael, & Snyder, 2006; Feldman & Dreher, 2012; King, 2001; Layous, Nelson, & Lyubomirsky, 2013; Marques, Lopez, & Pais-Ribeiro, 2011). Additionally, enhanced levels of hope may serve to buffer individuals against negative and stressful life events (Suldo & Michalowski, 2007).

Hope-based interventions have mostly conceptualised hope as a cognitive-motivational construct, as put forward by Snyder, Lopez, Shorey, Rand, and Feldman (2003). "The best possible future self" (King, 2001) and "goal mapping" (Feldman & Dreher, 2012) have been implemented successfully in adult, adolescent and child populations.

Against this backdrop, the aim of our study was to implement and evaluate the effect of a series of PPIs on the well-being and hope of a group of adolescents living in a CYCC. We conceptualised well-being broadly in terms of Keyes (2005) complete model of mental health, evaluating both the presence of positive mental health and the absence of depression and anxiety. Additionally, we viewed hope in cognitive-motivational terms, as described by Snyder et al. (2003). We expected that the PPIs would increase well-being and hope in this vulnerable population, whilst also decreasing depression and anxiety.

Method

Design

We implemented a non-randomised, quasi-experimental design. Due to the small sample size, we implemented non-random assignment to the experimental or control group. A matched-groups design was employed to control for potential selection effects (Morling, 2012). In order to mitigate potential third variable influences regarding age, gender and ethnicity, participants were allotted to the experimental or control group in as parsimonious a match as possible (Wilson & MacLean, 2011).

Participants

The sample comprised of 29 adolescents within a single care facility according to availability and accessibility. Their ages ranged between 14 and 18 years ($M = 16.31$; $SD = 1.37$). The majority of the sample were female (59%) and of African ethnicity (62%).

Procedure

The intervention was implemented by the first author and consisted of six structured, one-hour sessions conducted over a six-week period. The programme drew from research and interventions conducted by, among others, Suldo and Michalowski (2007), Proctor et al. (2011) and Feldman and Dreher (2012). Specific activities included counting blessings, engaging in acts of kindness, as well as identifying and using character strengths. To enhance hope, the "best possible self" (King, 2001) and goal mapping (Feldman & Dreher, 2012) exercises were implemented. In addition to these exercises, LKM and a savouring exercise were practiced in each session.

All but the first intervention session opened with a recapitulation of the previous week's intervention and a reflection on homework exercises if given (e.g. to continue counting blessings). The week's review was followed by a brief savouring exercise to facilitate focussing the group's attention to the present moment. Thereafter, group discussions centred on the specific topic (e.g. gratitude) took place. These discussions were followed by an activity associated with the topic of discussion. Each session closed with a five-minute LKM.

The two groups completed measures of well-being and hope before the study commenced, and again at one week after the conclusion of the intervention. Another follow-up measure was included five weeks later. The control group received the same intervention after collection of all data.

Ethical considerations

The Faculty of Humanities Research Ethics Committee provided ethical approval to conduct the research. The participating CYCCs Childcare Services Manager also provided permission to conduct the study. Further to this, the Childcare Services Manager acted as guardian to provide informed consent for each participant. Additionally, the participants provided assent to partake in the intervention. They were advised that all information provided would remain anonymous and confidential, that the study was not compulsory, and that they were allowed to withdraw from the programme at any time without any consequences.

In accordance with the principle of beneficence, every effort and intention was made to provide benefit to the adolescents involved in the study, moreover, each exercise and group session was considered with the core focus of doing no intentional or unintentional harm to the participants of the study.

Measures

Mental Health Continuum Short Form (MHC-SF; Keyes, 2009)

The MHC-SF measures positive mental health and consists of 14 items, comprising a three-factor structure measuring emotional, psychological and social well-being, yielding a total well-being score. Using a six-point Likert scale, the participants are asked to rate the frequency with which they experienced symptoms of positive mental health over the past month. Multiple studies have evidenced the MHC-SF to show high internal consistency ($a > .80$) among adults and adolescents (Keyes, 2005, 2006, 2009; Keyes et al., 2008; Westerhof & Keyes, 2010). Cronbach's alpha was also high in our study ($a = 86$).

Children's Hope Scale (CHS; Snyder et al., 1997)

The CHS is a six-item self-report questionnaire assessing dispositional hope among young people aged 8–16 years. The six items are divided equally to measure the bi-faceted construct of hope; the odd numbered items represent agentic thought, while the even numbered items represent pathways thinking. The CHS is hand-scored on a 6-point Likert scale, with total scores ranging between 6 (reflecting low levels of hope) and 36 (indicating high levels of hope). Snyder et al. (1997) reported that the CHS showed temporal stability, and established convergent, discriminant, predictive and incremental validity. The CHS also evidenced satisfactory psychometric properties among South African adolescents (Guse, De Bruin, & Kok, 2016). In our study, Cronbach's alpha was .71.

Revised Child Anxiety and Depression Scale – Short Version (RCADS-SV; Ebesutani et al., 2012)

The 25-item RCADS-SV is a brief measure of symptoms specific to pathological anxiety and depressive disorders in children and adolescents. The self-report questionnaire is scored on a 4-point Likert scale and is normed according to gender and grade level. Research revealed good internal consistency, with the anxiety subscale yielding Cronbach's alpha coefficients of .86 and .91 in non-referred and clinical samples, respectively. The depression subscale advanced Cronbach's alpha coefficients of .79 in the non-referred sample and .80 in the clinical sample. Additionally, good convergent and divergent, as well as acceptable concurrent validity was established with clinical diagnostic groups (Ebesutani et al., 2012). In the current study, the internal consistency of the RCAFS-SV was $a = .86$.

Data analysis

We employed descriptive statistics for the scale scores and demographics of the sample. The means of the dependent variables within and between the control and experimental groups before, as well as one and five weeks after the intervention, were compared using independent-samples t-tests (for between group comparisons) and paired-samples t-tests (for within-group comparisons). Parametric assumptions were tested beforehand in order to determine the appropriate tests of difference.

Results

The results of between and within-group analysis are presented in Tables 1 and 2 respectively. The baseline measures indicated that there were no statistically significant differences between the two groups on all measures before the intervention (Table 1). After the intervention, there was a slight decrease in hope in the experimental group and a marginal increase in the control group. However, these differences were not statistically significant. At the five-week follow-up, there were

Table 1. Differences between experimental and control groups pre, post and follow-up.

		Experimental group		Control group		
		Mean	SD	Mean	SD	p
CHS	Pre-test	25.86	5.57	21.87	5.81	.70
	Post-test	22.50	5.29	24.07	5.12	.42
	Follow-up	24.14	4.98	22.40	6.22	.41
MHC-SF	Pre-test	48.43	12.68	47.07	9.35	.74
	Post-test	48.21	11.81	47.60	8.39	.87
	Follow-up	48.71	14.71	43.80	14.53	.37
RCADS-SV	Pre-test	60.36	12.45	55.67	9.06	.25
	Post-test	56.71	13.36	54.93	10.74	.70
	Follow-up	55.29	11.53	58.07	18.15	.63

Note: MHC-SF: Mental Health Continuum – Short Form; RCADS-SV: Revised Child Anxiety and Depression Scale – Short Version.

Table 2. Differences within the experimental and control groups pre, post and follow-up.

		Pre-test		Post-test			Follow-up		
		Mean	SD	Mean	SD	p	Mean	SD	p
Experiment	CHS	25.86	5.57	22.5	5.29	.04	24.14	4.98	.22
	MHC-SF	48.43	12.68	48.21	11.81	.95	48.71	14.72	.94
	RCADS-SV	60.36	12.45	56.71	13.36	.29	55.29	11.53	.19
Control	CHS	21.87	5.81	24.07	5.12	.05	22.4	6.22	.65
	MHC-SF	47.07	9.35	47.60	8.39	.80	43.80	14.53	.24
	RCADS-SV	55.67	9.06	54.93	10.74	.77	58.07	18.15	.53

Note: MHC-SF: Mental Health Continuum – Short Form; RCADS-SV: Revised Child Anxiety and Depression Scale – Short Version.

no statistically significant differences in levels of hope between the two groups. Likewise, measures of the presence of positive functioning (MHC-SF) and absence of psychopathology (RCADS-SV) yielded no significant differences between the experimental and control groups one and five weeks after the intervention.

Within-group comparisons also showed no statistically significant changes in well-being and hope before and after the intervention, as reflected in Table 2. Although the experimental group scores showed a downward trend in depression and anxiety after the intervention, these results were not statistically significant when compared to baseline scores.

Discussion

The aim of our study was to examine the effect of a series of PPIs on well-being and hope of a group of adolescents living in a CYCC. Contrary to our expectations, the intervention did not lead to increased well-being or hope, nor was there a reduction in their levels of anxiety and depression. Our findings are similar to a few studies indicating that PPIs may not lead to improvements in well-being (Dickens, 2017; Marques et al., 2011) and hope (Weis & Speridakos, 2011) among youth, yet published research on non-significant findings of the effect of PPIs remain scarce. To gain a better understanding of these outcomes, we considered baseline levels of well-being and hope as well as moderating factors that may have contributed to the non-significant findings. We also offer a qualitative reflection on implementing PPIs.

The mean scores obtained on measures of well-being and hope were, surprisingly, comparable to, and generally better than those of referred and non-referred peer populations reported on in previous studies (see Gilman, Dooley, & Florell, 2006; Guse & Vermaak, 2011; McNeal et al., 2006; Snyder et al., 1997; van Schalkwyk & Wissing, 2010). This indicates that our sample evidenced higher levels of well-being and hope than expected. It is therefore possible that, with limited scope for improvement on subsequent measures, the ceiling effect could have contributed towards the non-significant outcomes of this study (Goodwin, 2010). Still, this explanation should be viewed with caution, as the objective reality of these youth may not warrant such high levels of well-being. However, a recent study among children in disadvantaged areas in the Western Cape, South Africa, also reported relatively high levels of subjective well-being (Savahl et al., 2015). It is evident that more research is needed regarding the measurement, experience and dynamics of well-being among South African children and adolescents.

Methodological and participatory moderators may also have played a role in the efficacy of the intervention. Nations and cultures may well differ in orientations of happiness, Park, Peterson, and Ruch (2009) reported that South Africans scored the highest out of 27 nations on an orientation towards pleasure. Thus, the MHC-SF, which largely incorporates elements of meaning and engagement in life, may not have fully tapped into the more transient, malleable and affect based facets of well-being, which may be relevant to this particular sample. Another methodological moderator that deserves mention is that of the duration of the intervention. In their meta-analysis of 51 interventions, Sin and Lyubomirsky (2009) reported that interventions of longer durations produced larger gains in well-being. The current study's six-week intervention fell within the second shortest of four temporal categories described in the meta-analysis. Therefore increasing the duration of the intervention may have yielded results of significance.

In addition to methodological moderators, research has repeatedly indicated individuals' behaviours, circumstances and characteristics may contribute to the efficacy of interventions. According to Lyubomirsky and colleagues (Lyubomirsky, Dickerhoof, Boehm, & Sheldon, 2011; Sin & Lyubomirsky, 2009; Sin, Della Porta, & Lyubomirsky, 2011), studies that recruited participants, as opposed to self-selection studies, yielded weaker and less robust effects. These scholars suggested that self-selection, driven by conscious knowledge of the purpose of interventions, was associated with motivation to achieve proposed outcomes. As such, motivation and increased effort placed on completing exercises, in combination with positive expectations, may have contributed towards stronger and more durable outcomes. Although participation in the current study was voluntary, adolescents did not personally pursue the programme, nor were they made aware of the aim to enhance levels of well-being and hope. As such, participants may have lacked motivation and effort in completing tasks.

Researchers have suggested that engaging in positive activities yields better results when those performing tasks receive social support and encouragement (Layous, Sheldon, & Lyubomirsky, 2014; Lyubomirsky et al., 2011; Sin et al., 2011). Research has indicated that children in care facilities experienced less emotional and social support than their family-nurtured counterparts (Allen & Vacca, 2010). Additionally, negative peer appraisals, increased loneliness, fewer friendships and less satisfaction within relationships are factors associated with cared-for youths (Dinisman, Montserrat, & Casas, 2012). It may, therefore, be possible that the vulnerable population in this study could have experienced lower levels of support, involvement, reassurance and encouragement when performing PPI exercises, which may have played a role in the effectiveness of the PPI.

Frequency and variation of practiced activities could impact on the effectiveness of gratitude and kindness interventions (Emmons & McCullough, 2003; Lyubomirsky, Sheldon, & Schkade, 2005; Sheldon & Lyubomirsky, 2012; Sheldon, Boehm, & Lyubomirsky, 2013; Tkach, 2005). Although instructions for these interventions were given in accordance with the findings, it appeared apparent in weekly feedback discussions that neither variety nor timing of tasks was maintained, even with repeated encouragement. Failure to observe task instructions may have resulted in participants performing tasks differently than suggested by empirical evidence, thereby contributing to the lack of change in outcome measures.

Research suggests that different positive activities may better suit participants with different personalities, values, strengths, interests, cultural preferences and circumstances (Lyubomirsky & Layous, 2013; Nelson & Lyubomirsky, 2012; Parks, Della Porta, Pierce, Zilca, & Lyubomirsky, 2012). Higher levels of person-activity fit have been associated with greater benefits (Nelson & Lyubomirsky, 2012; Schueller, 2011). The current study did not account for individual differences among participants, and therefore may have inadvertently employed activities not best suiting all participants, thus negatively influencing participant engagement and measured outcomes.

The statistically non-significant outcomes of our study may not be a true or complete reflection of the possible benefits of the intervention. Qualitatively we observed improvement in communication, confidence, self-esteem, positive affect and future-focussed optimism among some participants. Identifying and building character strengths noticeably enhanced levels of self-confidence and self-worth in a number of adolescents, with one participant privately reporting the impact these exercises had on assertiveness, self-belief and self-value. Additionally, several adolescents reported that practicing gratitude made them more aware of how much they had, as opposed to focussing on what was missing in their lives. Individuals mentioned that this practice, although "obvious", was not within their regular behaviour, and one went so far as to be grateful for the gratitude exercise, whilst another stated gratitude towards the intervention programme.

Youths also reflected positively on the weekly savouring activity, expressing gratitude for the immediate experience of indulgence as well as for their newfound capacity of identifying novel ways of appreciating regular experiences. Engaging in LKM reportedly brought calm into their otherwise hectic lives. A few adolescents remarked that LKM brought them a sense of connectedness and warmth with the prospect of offering anonymous help to others where they felt otherwise helpless.

Although participants seemed to enjoy the "acts of kindness" and "best possible future self" exercises on the whole, both activities generated a negative response in a few individuals. First, three members of the experimental group mentioned disliking performing acts of kindness as they found people either ridiculed them for their kind efforts or took advantage of their kindness. On the other hand, others expressed pleasure in performing kind acts, stating that the intrinsic reward of practicing kindness far outweighed the effort spent on the practice. Second, during the "best possible future self" exercise, one individual was reluctant to generate such an image to avoid the inevitable sadness of disappointment. According to this participant, not creating positive expectations was a protective mechanism developed following "a life of bad experiences". On the other hand, several adolescents in our study expressed vivid visualisations of seemingly achievable and desirable "best possible future self" images, once again exhibiting complete immersion and belief in the activity and in themselves. Finally, it was encouraging that a particularly troubled participant from the experimental group voluntarily attended the control group sessions as well, thereby experiencing the PPI for the second time. Additionally, an adolescent not involved in the study, and another who had declined consenting to participation, requested permission to attend the control group intervention, which attests to favourable reports from participants in the experimental group.

The subtleties in positive behavioural and attitudinal changes noted, as well as information regarding the benefits and drawbacks of specific activities implemented were not identified or assessed through the quantitative study. Including a qualitative component in future research on PPIs may, therefore, yield valuable information. Taking into account the elements discussed above, and the possible benefits vulnerable adolescents may gain from these activities, brief guidelines are offered for the implementation of PPIs among cared-for youth.

Future considerations for implementing PPIs in CYCCs

Considering that intervention duration moderates the effectiveness of PPIs with significantly greater benefit associated with longer treatments (Sin & Lyubomirsky, 2009), we suggest that PPIs longer than 8 weeks be implemented. Such practice may better support skill improvement and provide an opportunity for rehearsed positive activities to develop into habitual practices.

Where possible, intended intervention outcomes should be transparent, providing youths with explicit motive and intrinsic incentive to fully immerse themselves in the effortful practice of positive activities. Additionally, activities should be adapted to accommodate person-activity fit. For example, where acts of kindness may be better suited to more socially extroverted participants (Nelson & Lyubomirsky, 2012), youths in the current study who expressed discomfort with this exercise may have benefitted more by practicing self-compassion.

Bearing in mind the premise that expressing gratitude gives rise to an upward spiral of positive emotions, and that the participants seemed to enjoy the gratitude exercise, we suggest that exploring the virtue of gratitude, identifying blessings rather than burdens, and expressing appreciation for these blessings be the focus of the initial session. Similarly, participants provided positive feedback on engaging in LKM and savouring. It seems that these two activities served to focus the group's attention while simultaneously enhancing positive emotions in the present moment. We recommend that these exercises be repeated regularly throughout a PPI.

Finally, the PPIs included in this study were selected with specific focus on, and consideration of the characteristics, risk factors and vulnerabilities of adolescents residing in CYCCs. Based on our qualitative observations, PPIs aimed at optimising mental health and enhancing levels of well-being and hope in this population could include at least some of the activities implemented in this study.

In conclusion, despite the fact that our study did not find a significant improvement in well-being and hope for adolescents participating in a PPI, we expanded emerging literature that cautions against the over-optimistic implementation of positive interventions among youth. Qualitatively, however, there appeared to be value in these positive interventions for the adolescents in our sample and we encourage further mixed-method research on PPIs among vulnerable youth.

Disclosure statement

No potential conflict of interest was reported by the author(s).

Funding

This work was supported by National Research Foundation South Africa.

ORCID

Tharina Guse ⓘ http://orcid.org/0000-0001-9541-0392

References

Aguilar-Vafaie, M. E., Roshani, M., Hassanabadi, H., Masoudian, Z., & Afruz, G. A. (2011). Risk and protective factors for residential foster care adolescents. *Children and Youth Services Review, 33*(1), 1–15. doi:10.1016/j.childyouth.2010.08.005

Allen, B., & Vacca, J. S. (2010). Frequent moving has a negative affect on the school achievement of foster children makes the case for reform. *Children and Youth Services Review, 32*(6), 829–832. doi:10.1016/j.childyouth.2010.02.001

Berger, K. S. (2011). *The developing person through the lifespan* (8th ed.). New York, NY: Worth.

Bryant, F. B., Smart, C. M., & King, S. P. (2005). Using the past to enhance the present: Boosting happiness through positive reminiscence. *Journal of Happiness Studies, 6*(3), 227–260. doi:10.1007/s10902-005-3889-4

Bryant, F. B., & Veroff, J. (2007). *Savoring: A new model of positive experience.* Mahwah, NJ: Lawrence Erlbaum.

Cafasso, L. L. (1994). *Uplifts and hassles in the lives of young adolescents.* (Unpublished master's thesis). Chicago: Loyola University Chicago.

Cafasso, L. L. (1998). *Subjective well-being of inner-city resilient and non-resilient young adolescents.* (Unpublished doctoral dissertation). Chicago: Loyola University Chicago.

Cheavens, J. S., Feldman, D. B., Gum, A., Michael, S. T., & Snyder, C. R. (2006). Hope therapy in a community sample: A pilot investigation. *Social Indicators Research, 77*(1), 61–78.

Cicchetti, D. (2010). Resilience under conditions of extreme stress: A multilevel perspective. *World Psychiatry: Official Journal of the World Psychiatric Association (WPA), 9*(3), 145–154.

Cluver, L., & Gardner, F. (2007). Risk and protective factors for psychological well-being of children orphaned by AIDS in Cape Town: A qualitative study of children and caregivers' perspectives. *AIDS Care, 19*(3), 318–325. doi:10.1080/09540120600986578

Coleman, J., & Hagell, A. (2007). *Adolescence, risk and resilience: Against the odds.* Chichester: John Wiley.

Daniel, B., Wassell, S., & Gilligan, R. (1999). It's just common sense isn't it?': Exploring ways of putting the theory of resilience into action. *Adoption & Fostering, 23*(3), 6–15. doi:10.1177/030857599902300303

Dickens, L. R. (2017). Using gratitude to promote positive change: A series of meta-analyses investigating the effectiveness of gratitude interventions. *Basic and Applied Social Psychology, 39*(4), 193–208.

Dinisman, T., Montserrat, C., & Casas, F. (2012). The subjective well-being of Spanish adolescents: Variations according to different living arrangements. *Children and Youth Services Review, 34*(12), 2374–2380. doi:10.1016/j.childyouth.2012.09.005

Ebesutani, C., Reise, S. P., Chorpita, B. F., Ale, C., Regan, J., Young, J., … Weisz, J. R. (2012). The revised child anxiety and depression scale-short version: Scale reduction via exploratory bifactor modeling of the broad anxiety factor. *Psychological Assessment, 24*(4), 833–845. doi:10.1037/a0027283

Emmons, R. A., & McCullough, M. E. (2003). Counting blessings versus burdens: An experimental investigation of gratitude and subjective well-being in daily life. *Journal of Personality and Social Psychology, 84*(2), 377–389. doi:10.1037/0022-3514.84.2.377

Epstein, M. H. (2000). The behavioral and emotional rating scale: A strength-based approach to assessment. *Assessment for Effective Intervention, 25*(3), 249–256. doi:10.1177/073724770002500304

Evans, D. L., Foa, E. B., Gur, R. E., Hendin, R., O'Brien, C. P., Seligman, M. E. P., & Walsh, B. T. (Eds.). (2005). *Treating and preventing adolescent mental health disorders: What we know and what we don't know. A research agenda for improving the mental health of our youth.* New York, NY: Oxford University Press.

Feldman, D. B., & Dreher, D. E. (2012). Can hope be changed in 90 minutes? Testing the efficacy of a single-session goal-pursuit intervention for college students. *Journal of Happiness Studies, 13*(4), 745–759. doi:10.1007/s10902-011-9292-4

Fergusson, D. M., & Horwood, L. J. (2003). Resilience to childhood adversity: Results of a 21-year study. In S. S. Luthar (Ed.), *Resilience and vulnerablity: Adaptation in the context of childhood adversities* (pp. 130–155). Cambridge: Cambridge University Press.

Fredrickson, B. L. (2001). The role of positive emotions in positive psychology: The broaden-and-build theory of positive emotions.. *American Psychologist, 56*(3), 218–226.

Fredrickson, B. L., Cohn, M. A., Coffey, K. A., Pek, J., & Finkel, S. M. (2008). Open hearts build lives: Positive emotions, induced through loving-kindness meditation, build consequential personal resources. *Journal of Personality and Social Psychology, 95*(5), 1045–1062. doi:10.1037/a0013262

Froh, J. J., Sefick, W. J., & Emmons, R. A. (2008). Counting blessings in early adolescents: An experimental study of gratitude and subjective well-being. *Journal of School Psychology, 46*(2), 213–233. doi:10.1016/j.jsp.2007.03.005

Gander, F., Proyer, R., Ruch, W., & Wyss, T. (2012). Strength-based positive interventions: Further evidence for their potential in enhancing well-being and alleviating depression. *Journal of Happiness Studies.* doi:10.1007/s10902-012-9380-0

Garland, E. L., Fredrickson, B., Kring, A. M., Johnson, D. P., Meyer, P. S., & Penn, D. L. (2010). Upward spirals of positive emotions counter downward spirals of negativity: Insights from the broaden-and-build theory and affective neuroscience on the treatment of emotion dysfunctions and deficits in psychopathology. *Clinical Psychology Review, 30*(7), 849–864. doi:10.1016/j.cpr.2010.03.002

Gilman, R., Dooley, J., & Florell, D. (2006). Relative levels of hope and their relationship with academic and psychological indicators among adolescents. *Journal of Social and Clinical Psychology, 25*(2), 166–178.

Goodwin, C. J. (2010). *Research in psychology: Methods and design* (6th ed.). Hoboken, NJ: Wiley.

Guse, T. (2014). Activities and programmes to enhance well-being. In M. P. Wissing, J. Potgieter, T. Guse, I. Khumalo, & L. Nel (Eds.), *Towards flourishing: Contextualizing positive psychology.* Pretoria: Van Schaik.

Guse, T., De Bruin, G. P., & Kok, M. (2016). Validation of the children's hope scale in a sample of South African adolescents. *Child Indicators Research, 9*(3), 757–770.

Guse, T., & Vermaak, Y. (2011). Hope, psychosocial well-being and socioeconomic status among a group of South African adolescents. *Journal of Psychology in Africa, 21*(4), 527–533.

Hagen, K. B., Myers, B. J., & Mackintosh, V. H. (2005). Hope, social support, and behavioral problems in at-risk children. *American Journal of Orthopsychiatry, 75*(2), 211–219.

Herth, K. (1998). Hope as seen through the eyes of homeless children. *Journal of Advanced Nursing, 28*(5), 1053–1062. Retrieved from http://www.ncbi.nlm.nih.gov/pubmed/9840877

Hurley, D. B., & Kwon, P. (2012). Results of a study to increase savoring the moment: Differential impact on positive and negative outcomes. *Journal of Happiness Studies, 13*(4), 579–588.

Hutcherson, C. A., Seppala, E. M., & Gross, J. J. (2008). Loving-kindness meditation increases social connectedness. *Emotion, 8*(5), 720–724. doi:10.1037/a0013237

Jose, P. E., Lim, B. T., & Bryant, F. B. (2012). Does savoring increase happiness? A daily diary study. *The Journal of Positive Psychology, 7*(3), 176–187.

Keyes, C. L. M. (2005). The subjective well-being of America's youth: Toward a comprehensive assessment. *Adolescent and Family Health, 4*, 3–11.

Keyes, C. L. M. (2006). Mental health in adolescence: Is America's youth flourishing? *American Journal of Orthopsychiatry, 76*(3), 395–402.

Keyes, C. L. M. (2009). Atlanta: *Brief description of the mental health continuum short form (MHC-SF)*. Retrieved from http://www.sociology.emory.edu/ckeyes/

Keyes, C. L. M., Wissing, M., Potgieter, J. P., Temane, M., Kruger, A., & van Rooy, S. (2008). Evaluation of the Mental Health Continuum – Short Form (MHC – SF) in Setswana-speaking South Africans. *Clinical Psychology and Psychotherapy, 15* (3), 181–192.

King, L. (2001). The health benefits of writing about life goals. *Personality and Social Psychology Bulletin, 27*(7), 798–807. doi:10.1177/0146167201277003

Kjelsberg, E., & Nygren, P. (2004). The prevalence of emotional and behavioural problems in institutionalized childcare clients. *Nordic Journal of Psychiatry, 58*(4), 319–325. doi:10.1080/08039480410005846

Kristeller, J. L., & Johnson, T. (2005). Cultivating loving kindness: A two-stage model of the effects of meditation on empathy, compassion, and altruism. *Zygon Journal of Religion and Science, 40*(2), 391–408.

Layous, K., Lee, H., Choi, I., & Lyubomirsky, S. (2013). Culture matters when designing a successful happiness-increasing activity. *Journal of Cross-Cultural Psychology, 44*, 1294–1303.

Layous, K., Nelson, K. S., & Lyubomirsky, S. (2013). What is the optimal way to deliver a positive activity intervention? The case of writing about one's best possible selves. *Journal of Happiness Studies, 14*(2), 635–654. doi:10.1007/s10902-012-9346-2

Layous, K., Sheldon, K. M., & Lyubomirsky, S. (2014). The prospects, practices, and prescriptions for the pursuit of happiness. In S. Joseph (Ed.), *Positive psychology in practice* (2nd ed., pp. 248–273). New York, NY: John Wiley.

Leslie, L. K., Gordon, J. N., Ganger, W., & Gist, K. (2002). Developmental delay in young children in child welfare by initial placement type. *Infant Mental Health Journal, 23*(5), 496–516.

Lyubomirsky, S., Dickerhoof, R., Boehm, J. K., & Sheldon, K. M. (2011). Becoming happier takes both a will and a proper way: An experimental longitudinal intervention to boost well-being. *Emotion, 11*(2), 391–402. doi:10.1037/a0022575

Lyubomirsky, S., & Layous, K. (2013). How do simple positive activities increase well-being? *Current Directions in Psychological Science, 22*(1), 57–62. doi:10.1177/0963721412469809

Lyubomirsky, S., Sheldon, K. M., & Schkade, D. (2005). Pursuing happiness: The architecture of sustainable change. *Review of General Psychology, 9*(2), 111–131. doi:10.1037/1089-2680.9.2.111

Marques, S. C., Lopez, S. J., & Pais-Ribeiro, J. (2011). "Building hope for the future": A program to foster strengths in middle-school students. *Journal of Happiness Studies, 12*(1), 139–152. doi:10.1007/s10902-009-9180-3

McNeal, R., Handwerk, M. L., Field, C. E., Roberts, M. C., Soper, S., Huefner, J. C., & Ringle, J. L. (2006). Hope as an outcome variable among youths in a residential care setting. *American Journal of Orthopsychiatry, 76*(3), 304–311. doi:10.1037/0002-9432.76.3.304

Milkman, H., & Wanberg, K. (2012). Adolescent development and pathways to problem behavior. In *Criminal conduct and substance abuse treatment for adolescents: Pathways to self-discovery and change* (2nd ed., pp. 21–54). Washington, DC: Sage.

Moore, W. (2010). *Investigation of character strengths among college attendees with and without a history of child abuse* (Unpublished doctoral dissertation). Massachusetts School of Professional Psychology, Massachusetts.

Morling, B. (2012). *Research methods in psychology: Evaluating a world of information*. New York, NY: W.W. Norton.

Mullan, C., & Fitzsimons, L. (2006). *The mental health of looked after children/care leavers in Northern Ireland: A literature review* (Research Report). Retrieved from http://www.voypic.org/publications/research-policy-reports

Nelson, S. K., & Lyubomirsky, S. (2012). Finding happiness: Tailoring positive activities for optimal well-being benefits. In M. Tugade, M. Shiota, & L. Kirby (Eds.), *Handbook of positive emotions* (pp. 275–293). New York, NY: Guilford.

Park, N. (2004). Character strengths and positive youth development. *The Annals of the American Academy of Political and Social Science, 591*(1), 40–54. doi:10.1177/0002716203260079

Park, N., Peterson, C., & Ruch, W. (2009). Orientations to happiness and life satisfaction in twenty-seven nations. *The Journal of Positive Psychology, 4*(4), 273–279. doi:10.1080/17439760902933690

Park, N., Peterson, C., & Seligman, M. E. (2004). Strengths of character and well-being. *Journal of Social and Clinical Psychology, 23*(5), 603–619.

Parks, A., & Biswas-Diener, R. (2013). Positive interventions: Past, present and future. In T. B. Kashdan, & J. Ciarrochi (Eds.), *The context press mindfulness and acceptance practica series. Mindfulness, acceptance, and positive psychology: The seven foundations of well-being* (pp. 140–165). Oakland, CA: Context Press/New Harbinger.

Parks, A. C., Della Porta, M. D., Pierce, R. S., Zilca, R., & Lyubomirsky, S. (2012). Pursuing happiness in everyday life: The characteristics and behaviors of online happiness seekers. *Emotion, 12*(6), 1222–1234. doi:10.1037/a0028587

Peterson, C., & Seligman, M. E. P. (2004). *Character strengths and virtues: A classification and handbook.* Washington, DC: American Psychological Association.

Proctor, C., Tsukayama, E., Wood, A. M., Maltby, J., Eades, J., & Linley, P. A. (2011). Strengths gym: The impact of a character strengths-based intervention on the life satisfaction and well-being of adolescents. *The Journal of Positive Psychology, 6*(5), 377–388. doi:10.1080/17439760.2011.594079

Quinlan, D., Swain, N., & Vella-Brodrick, D. A. (2012). Character strengths interventions: Building on what we know for improved outcomes. *Journal of Happiness Studies, 13*(6), 1145–1163.

Richardson, J., & Lelliott, P. (2003). Mental health of looked after children. *Advances in Psychiatric Treatment, 9*(4), 249–256. doi:10.1192/apt.9.4.249

Santrock, J. W. (2010). *Adolescence* (13th ed.). New York, NY: McGraw Hill.

Savahl, S., Adams, S., Isaacs, S., September, R., Hendricks, G., & Noordien, Z. (2015). Subjective well-being amongst a sample of South African children: A descriptive study. *Child Indicators Research, 8*(1), 211–226.

Schueller, S. M. (2011). To each his own well-being boosting intervention: Using preference to guide selection. *The Journal of Positive Psychology, 6*(4), 300–313.

Seligman, M. E. P. (2002). Positive psychology, positive prevention, and positive therapy. In C. R. Snyder & S. J. Lopez (Eds.), *Handbook of positive psychology* (pp. 3–9). Oxford: Oxford University Press.

Seligman, M. E. P., & Csikszentmihalyi, M. (2000). Positive psychology: An introduction. *American Psychologist, 55*(1), 5–14.

Seligman, M. E. P., Steen, T. A., Park, N., & Peterson, C. (2005). Positive psychology progress: Empirical validation of interventions. *American Psychologist, 60*(5), 410–421. doi:10.1037/0003-066X.60.5.410

Sheldon, K. M., Boehm, J. K., & Lyubomirsky, S. (2013). Variety is the spice of happiness: The hedonic adaptation prevention model. In I. Boniwell, & S. David (Eds.), *Oxford handbook of happiness* (pp. 901–915). Oxford: Oxford University Press.

Sheldon, K. M., & Lyubomirsky, S. (2012). The challenge of staying happier. *Personality and Social Psychology Bulletin, 38*, 670–680.

Sin, N. L., Della Porta, M. D., & Lyubomirsky, S. (2011). Tailoring positive psychology interventions to treat depressed individuals. In S. I. Donaldson, M. Csikszentmihalyi, & J. Nakamura (Eds.), *Applied positive psychology: Improving everyday life, health, schools, work and society* (pp. 79–96). New York, NY: Routledge.

Sin, N. L., & Lyubomirsky, S. (2009). Enhancing well-being and alleviating depressive symptoms with positive psychology interventions: A practice-friendly meta-analysis. *Journal of Clinical Psychology, 65*(5), 467–487.

Snyder, C. R., Hoza, B., Pelham, W. E., Rapoff, M., Ware, L., Danovsky, M., … Stahl, K. (1997). The development and validation of the Children's Hope Scale. *Journal of Pediatric Psychology, 22*(3), 399–421.

Snyder, C. R., Lopez, S., Shorey, H., Rand, K., & Feldman, D. (2003). Hope theory, measurements, and applications to school psychology. *School Psychology Quarterly, 18*, 122–139.

Suldo, S. M., & Michalowski, J. A. (2007). *Subjective well-being intervention program [Procedures manual].* Retrieved from http://www.coedu.usf.edu/schoolpsych/ Faculty/Suldo.html

Tkach, C. (2005). *Unlocking the treasury of human kindness: Enduring improvements in mood, happiness, and self-evaluations* (Unpublished doctoral dissertation). Riverside, CA: University of California.

van Schalkwyk, I., & Wissing, M. P. (2010). Psychosocial well-being in a group of South African adolescents. *Journal of Psychology in Africa, 20*(1), 53–60.

Weis, R., & Speridakos, E. C. (2011). A meta-analysis of hope enhancement strategies in clinical and community settings. *Psychology of Well-Being: Theory, Research and Practice, 1*(1), 5–16.

Westerhof, G. J., & Keyes, C. L. M. (2010). Mental illness and mental health: The two continua model across the lifespan. *Journal of Adult Development, 17*(2), 110–119. doi:10.1007/s10804-009-9082-y

Wilson, S., & MacLean, R. (2011). *Research methods and data analysis for psychology.* London: McGraw-Hill.

Zimmer, M. H., & Panko, L. M. (2006). Developmental status and service use among children in the child welfare system – A national survey. *Archives of Pediatrics & Adolescent Medicine, 160*(2), 183–188.

The broader aims of career development: mental health, wellbeing and work

Dave E. Redekopp and Michael Huston

ABSTRACT
Work is a significant factor in mental health and wellbeing outcomes. Career development processes can be helpful in finding and managing work trajectories that lead to these as well as additional wellbeing outcomes. Conversely, mental illness can impede the acquisition and retention of suitable work as well as the ability to fully engage in career development. Evidence for the interactive relationships between work, career development, mental health and mental illness is reviewed, with an emphasis on the relationships between work (both "good" and "bad") and wellbeing outcomes. Evidence for counselling/guidance interventions, organisational interventions and policy directions follow, concluding with suggestions for improving wellbeing via career development interventions.

Introduction

> If a young man chooses his vocation so that his best abilities and enthusiasms will be united with his daily work, he has laid the foundations of success and happiness … and if his occupation is merely a means of making a living, and the work he loves to do is side-tracked into the evening hours, or pushed out of his life altogether, he will be only a fraction of the man he ought to be. (Parsons, 1909, p. 10)

Philosophers (e.g. Marx, 1891), sociologists (e.g. Schumpeter, 1942), psychologists (e.g. Adler, 1958) and economists (e.g. Layard, 2011) have reflected on the work-wellbeing relationship before and after Parsons' tribute to intentionally choosing a vocation. The career development field's attention to the relationship, however, has waxed and waned over the last century. This wavering focus is curious, given that the justification for career development practices rests largely on a strong, positive and causal relationship linking career development practice and subsequent work choice with happiness, wellbeing and quality of life. Perhaps occasional lapses of visible interest have been due to simply assuming that there is a strong relationship between work and wellbeing (Super, 1957), the difficulty in establishing evidence for this relationship, and/or placing a priority on creating career development theories and practices over showing the value of these theories and practices.

Regardless of the reasons, "Career development has been slow to take seriously its whole mandate" (Peruniak, 2010, p. x), a situation that appears to be rapidly changing. Conceptual seeds urging a focus on work-wellbeing relationships planted by authors such as Brown and Brooks (1985), Herr (1989), and Waddell and Burton (2006) have germinated in the past few years and are quickly spreading roots nourished by career development thinkers such as Blustein (Blustein, 2006, 2008), McIlveen (Kossen & McIlveen, 2017) , Peruniak (2010) and (Robertson, 2013b) as well as by researchers in related fields such as mental health promotion (Czabała & Charzyńska, 2014), medicine

(Waddell & Burton, 2006), occupational health and safety (Warr & Inceoglu, 2018) and sociology (Brand, 2015).

Reviewing the recent work in this area quickly highlights that career development and mental health are related. The work also reveals that a number of distinct concepts are, or are prone to being, conflated particularly by the public and non-academic practitioners who do not have the time or interest to dive into the details of these connections. Given the seemingly broad interest in and proliferation of work on career development and mental health relationships, now is a good time to begin making clear distinctions between constructs.[1]

The first distinction lies between two terms that often blend together in the layperson's mind: "career development" and "work." This blending is regrettable but understandable in public discourse. We see it, however, seep into the minds of educators, psychologists/therapists, human resources professionals and others, all of whom can have a significant impact on individuals' career development and their work. In this paper, we will differentiate "career development" as any learning, maturation or growth related to the preparation for, adaptation to, management of and movement between life roles. The role of worker is a significant role that is historically the primary focus for career development theories. "Work" is an overlapping construct that can refer to paid or unpaid sets of goal-oriented activities but, for our purposes, will refer to remunerated sets of activities typically done for an organisation in order to earn a living. We clarify this distinction because career development can influence mental health whether or not work is involved, and work can influence mental health whether or not any intentional career development has occurred.

The second distinction is between "mental illness" and "mental health." Although historically viewed as two ends of a continuum or two distinct states, support can be garnered for seeing them as distinct but related concepts (Huppert, 2014). We draw upon Keyes' (2005) 2-continua model, which conceives low to high mental illness forming one continuum (x-axis) and low to high mental health forming another (y-axis), resulting in the possibility of varying combinations of each (e.g. high mental health and high mental illness, low mental health and low mental illness). In the 2-continua model, it is possible to have low mental health ("languishing" is Keyes' term for this) without having mental illness, just as it is possible to possess mental health ("flourishing", in Keyes' model) even with a mental illness. We employ the 2-continua model for both conceptual and selfish reasons. Conceptually, the distinction has merit and is supported by evidence (Westerhof & Keyes, 2010). Further, mental illness takes a myriad of forms (American Psychiatric Association, 2013), each of which may interact differently with work and career development. The varieties of mental health, if there are varieties, have not yet been identified, leaving mental health as a global construct. Selfishly, it is far easier to show the positive effects of both career development and work on mental health (which, as career development specialists, serves our interests) than it is to demonstrate positive effects of either on mental illness.

We are working on a larger framework that will delineate the reciprocal relationships between work and mental illness, work and mental health, career development and mental illness, and career development and mental health. In this paper, our focus is on work and mental health and work and mental illness. We use the term "wellbeing" in a global manner aligned with Robertson (2013b): i.e. positive or high wellbeing is the condition of low mental illness, high mental health and high physical health. In this paper, wellbeing captures both mental health and mental illness.

Mental illness and work

Mental illness is the leading cause both of work absence due to sickness and long-term disability (Petrie et al., 2018), with depression and anxiety accounting for the majority (Harvey et al., 2013). Costs to organisations and national economies are considerable. Further, absenteeism and lost productivity due to presenteeism (working while sick) is a well-recognized concern (Harvey et al., 2013; World Health Organization, 2017).

Effect(s) of mental illness on work

Mental illness in almost any form impacts the way individuals enter and participate in work. Earlier age of onset, increased severity, and increased duration of illness episodes are predictive of increasingly deleterious work effects. Individuals with permanent, severe mental illnesses are more greatly impacted because the development of both their sense of self-efficacy in areas related to work and their identity as a worker are hindered. Opportunities to explore and to test one's aptitudes and abilities related to work are key in the development of work identity and self-efficacy. Severe and permanent mental illness can prevent entry into and participation in working life. With regards to individuals with severe mental illness, Waddell and Burton (2006) found that work is neither harmful nor helpful regarding their particular illness, but it appears that overall wellbeing is improved by work.

Severe mental illness is less common and more of a hinderance to entry and participation in work than most mental illness, which can be characterised as episodic, mild to moderate, and usually starting later in life (e.g. after age 25). Severe mental illness tends to interfere more with work, working life, and organisations than more common mental illness. Anxiety, depression and substance use disorders are the most common mental illness concerns that interfere with working life (Harvey et al., 2013; Joyce et al., 2016) and are largely responsible for the sickness absence costs posed to organisations and economies in most developed countries. The costs attributable to the consequences of mental illness for work include: absenteeism (due to increase sickness, poor health, physical maladies such as back pain), presenteeism (working while incapacitated due to illness), poor work performance (reduced productivity, accidents, impaired decision-making), poor attitude (low motivation and commitment, ineffectiveness, poor time management), poor relationships (increased tension between colleagues, poor relationships with clients) and increased disciplinary problems (Harnois & Gabriel, 2000; Sainsbury et al., 2008).

Work's role in the development of mental illness

Of the global population, 4.4% experience depression and 3.6% experience anxiety, with slightly greater representation in women than in men, and more (80%) in low and middle-income than high income countries (World Health Organization, 2017). Exposure to stressors such as job loss, unemployment, and other losses also increases the likelihood of experiencing one of these mental health concerns. Work itself can also be a contributing factor. In a meta-review on the topic, Harvey et al. (2017) identified work factors contributing to the development of mental health concerns at work. Although available evidence prevents attributions about causality, the work problems (stressors) associated with mental health concerns include:

- High job strain (excessive demands and limited autonomy)
- Lack of appropriate reward for effort (seemingly unfair pay and/or recognition)
- Lack of organisational justice (referring to the way information and resources flow within an organisation)
- Job insecurity and downsizing (a significant stressor frequently associated with anxiety and depressions which can persist after the causal event is resolved, even for those who keep their jobs)
- Interpersonal problems and being bullied

Career development and mental health & mental illness

Although this paper's focus is on work and wellbeing, a brief synopsis of some key relationships between career development and wellbeing are provided here in anticipation of the practice implications of the work-wellbeing connections.

Mental illness can variously affect a host of career development factors, including individuals' perceptions of barriers to employment, actual barriers to employment, motivation, anticipated career-related outcomes, self-confidence, exploratory behaviour and identity formation (Boychuk, Lysaght, & Stuart, 2018; Luciano & Carpenter-Song, 2014) as well as abilities to effectively make use of counselling or guidance services (Caporoso & Kiselica, 2004).

There is little direct evidence supporting a positive relationship between career intervention and the reduction of mental illness symptoms. Literature in this area tends to report on case studies or address indirect effects (Robertson, 2013b). The indirect effects typically feature mental health indicators rather than focusing on mental illness symptoms. For example, Robertson selects self-efficacy as a robust construct for which there is considerable evidence for its development via career interventions. Self-efficacy is related to wellbeing, and therefore career interventions can be seen to promote wellbeing. Similarly, Robertson summarises arguments that career interventions enhance optimism, reframe failure as possible assets, develop social identity, reduce career anxiety and reduce cognitive dissonance, each of which contributes to wellbeing.

Work's effects on mental health/wellbeing

Comprehensive reviews of research findings (Brand, 2015; Lindberg & Vingård, 2012; Robertson, 2013b; Waddell & Burton, 2006), constructs (Huppert, 2014) and practices (Czabała, Charzyńska, & Mroziak, 2011) reveal numerous robust relationships between wellbeing and working, worklessness, threats of worklessness, losing work, obtaining work and quality of work. Waddell and Burton's (2006) review of studies on work and health found that "work and paid employment are generally beneficial for physical and mental health and well-being" (p. 10), recognising that job quality and social context play a significant role. Also, unemployment is associated with numerous health problems, including poorer wellbeing, greater psychological distress, increased minor mental illness and increased parasuicide, with strong evidence that unemployment is a causal and/or contributing factor. Waddell & Burton note that most of the reviewed research focused on men, but that the same effects appear to hold true for women. They also reported on age-specific findings, noting that unemployed young people have, in comparison to employed young people, higher mortality rates, largely due to accidents and suicide, as well as slightly more moderate effects as listed above.

Regarding re-employment (which, in Waddell and Burton's (2006) review, includes school leavers entering employment or further training/education), school leavers who enter poor employment can suffer a decline in wellbeing, whereas those who enter satisfactory employment or training experience improvements in both physical and psychological symptoms compared with those entering unemployment. Re-employment for unemployed adults is associated with a variety of wellbeing improvements, such as self-esteem and self-satisfaction as well as less psychological distress and fewer mental illness diagnoses. As with the other findings described above, "health benefits depend on the job or the training being 'satisfactory' while 'unsatisfactory' jobs may be little better than unemployment" (p. 19).

Brand's (2015) review found that losing work (an event, job loss, that is differentiated from unemployment, a state) is associated with significant long-term earnings loss (approximately 20% lifetime earnings loss), increased likelihood of obtaining part-time rather than full-time work after job loss, greater job instability and increased likelihood of lower quality jobs. These negative effects are stronger with older workers and with less educated workers, according to Brand. Wellbeing effects of job loss include "higher levels of depressive symptoms, somatisation, anxiety, and the loss of psychosocial assets" (p. 365), adjusted for baseline psychological health (Brand notes that research has not connected job loss with clinically diagnosable depression and anxiety, but self-reports show 15%–30% increases when compared to nondisplaced workers). Wellbeing effects vary by history (the first job loss may have greater impact than the second or third), occupational level (professionals may feel more shock than less-skilled workers), commitment (workers whose sense of self is strongly connected to work suffer more), and context (wellbeing may suffer less if job loss is due to external factors rather than their performance. Brand also describes the effects of job loss on physical well-being, citing numerous ways in which health worsens, ranging from substantial increase (50%–100%) in mortality within a year of job loss to increases in self-destructive behaviours and suicide. Job-loss's relationships to families (e.g. increased family tension, increased marital dissolution) and communities (e.g. decreased community group involvement, decreased social engagement) are also noted by Brand.

The findings above distinguish between "satisfactory" or "good" and "unsatisfactory" or "bad" work. A key outcome of "good" work, job satisfaction, is known to be quite strongly and positively correlated with health, including mental health (Faragher, Cass, & Cooper, 2005). The actual characteristics of "good" work, however, are not clear. Lindberg and Vingård (2012) review of 152 publications on healthy work environments found only one study on the indicators of healthy work environments or "satisfactory" work. The remaining publications addressed what experts believed and/or employees felt should contribute to healthy work environments. Compiling these findings resulted in nine "most pronounced components ... : collaboration/teamwork; growth and development of the individual; recognition; employee involvement; positive, accessible and fair leader; autonomy and empowerment; appropriate staffing; skilled communication; and safe physical work" (p. 3036–3037). In a more recent, preliminary review, Lindberg, Karlsson, Nordlöf, Engström, and Vingård (2017) identified a list of 12 most frequently researched factors:

> style of leadership; empowerment, autonomy, control at work, participation; possibilities for own development; positive work climate, social support from supervisor, communication supervisor-employee; clear goals; appreciation from supervisors, colleagues, customers; work time control, enough time; effort-reward balance; intellectually stimulating; job security ("Results", para. 3).

A meta-review of studies on the mental health benefits of employment (Modini et al., 2016) summarised that the factors of "good" work are unclear and the contributions and interactions of possible factors on mental health are not known.

Counsellors will notice the absence in this list of "use of personal qualities", "in line with personal values", "work-life balance" and "benefit to society", issues that likely emerge frequently in guidance and counselling sessions. Lindberg and Vingård's (2012) review includes these factors but, with the exception of "in line with personal values" (which was included in four studies), found only one study that had addressed each factor. As seen in Lindberg et al.'s (2017) review, these factors may not appear as top factors simply because they have not been attended to in the research. Since Lindberg & Vingård's review, (Praskova, Creed, & Hood, 2015) found positive relationships between "career calling" and life satisfaction, work effort, greater use of career strategies, and higher emotional regulation. Robertson's (2013b) review indicates that congruence of interests and work is known to be associated, albeit only moderately, with job satisfaction, and that job satisfaction is related to health and life satisfaction. Clearly, more research is needed to clarify the above relationships, the characteristics of "good" work and ways to promote these characteristics. As Huppert (2014) points out, half of one's life is spent at work and we would do well by making work health-promoting, especially since there is evidence indicating that workplace programmes encourage mental health spillover into family and other relationships.

Another set of work-loss variables connected with wellbeing outcomes relates to the uncertainty of work. It would not be surprising to find that worries about gaining or losing full-time employment or about successfully transitioning from one temporary contract or gig to another exacted a wellbeing toll. Conversely, individuals with mental health concerns may be more susceptible to losing work or being engaged in contract work. The evidence is mixed, generally supporting the idea that worrying about obtaining work is anxiety producing, particularly with youth entering the labour market (Thern, de Munter, Hemmingsson, & Rasmussen, 2017). For example, a survey of youth aged 18–24 in Canada found 90% of respondents reporting uncomfortable levels of stress and, of those, 86% attributed stress to underemployment (Sun Life Financial, 2012). Thern et al.

note that the deleterious effects of this anxiety remain with youth far longer than with individuals who have been in the workforce for some time.

In a comprehensive study of longitudinal data involving 8,000 individuals over 17 years, Dawson, Veliziotis, Pacheco, and Webber (2015) found that individuals with mental health concerns are more likely than those without to transition from permanent to temporary employment: i.e. mental health problems cause the move to precarious employment (they have no evidence whether this move is by volition or force). They recognise that the literature is divided on the issue of causality and recommend further, cause-focussed research. This research will become particularly important as the economy increasingly involves "gig" work.

Summary: work's relationships with mental illness, mental health and wellbeing

In all of the research reviewed above, there is little certainty about the degree of most findings (e.g. How little work is enough work to see benefit? How much is too much?) (Waddell & Burton, 2006), the details within the constructs (e.g. what are the key differentiators between "good" and "bad" work?) and the direction of effects (e.g. is temporary work more available to individuals with mental illness or is it a cause of a mental illness?). However, several consistent findings point to career interventions likely to promote happiness and wellbeing in the workplace:

- Generally, work has positive effects on mental health and wellbeing. For most people in most situations, working is better than not working. This point is particularly evident through consideration of the experience of unemployed versus employed persons. Work can be a source of identity, social support, financial and other coping resources, and it can also bolster one's sense of purpose.
- Person-work fit is related to mental health/wellbeing. For psychological health, working is better than not working, and employment that fits (work that affords an expression of one's interests, values, strengths and needs in work or in other life roles) is related to better wellbeing outcomes than work that does not fit.
- Work-related factors can play a role in the development of mental illness. Excessive job strain, related stress, role ambiguity, job insecurity, unsupportive management, and psychosocial concerns are associated with sickness absence and presenteeism due to mental health concerns. It is likely that certain work contexts, environments, and factors have a causal role in the development of mental illness.
- Work-related factors can contribute to the development or maintenance of mental health/wellbeing and can be effective adjuncts in the treatment of a range mental illness concerns. Fitting work in good work environments bolsters mental health and wellbeing.

Career interventions to promote happiness and wellbeing

Individuals' wellbeing would be well-served by career development guidance and counselling services that helped them, in priority, choose and find work, find work that fits, find fitting work in the healthiest workplaces possible (i.e. finding "good work"), and develop abilities to continuously adapt to changing circumstances. Here, we address guidance and counselling interventions, organisational interventions and policy approaches that lead to finding and/or creating "good" work or, more specifically, a work environment that is conducive to mental health and wellbeing.

Guidance and counselling interventions

The belief that career intervention plays a significant role in mental health outcomes is longstanding (e.g. Brown & Brooks, 1985; Herr, 1989) but direct evidence linking career intervention to mental health outcomes is limited. Some recent studies make this link. For example, career management interventions leading to career preparedness (bolstering career self-efficacy and preparation against setbacks) can significantly reduce symptoms of depression (Ahola, Vuori, Toppinen-Tanner, Mutanen, & Honkonen, 2012; McGonagle, Beatty, & Joffe, 2014; Salmela-Aro, Mutanen, & Vuori, 2012; Vuori, Toppinen-Tanner, & Mutanen, 2012) and longer sickness absence (Toppinen-Tanner, Böckerman, Mutanen, Martimo, & Vuori, 2016). The mechanisms at play are unclear. For most people, work is central to their identity, meaningful, and a focal point of their purpose and sense of connection with others and the outer world, leading some researchers to hypothesise that coping more effectively with these meaningful concerns is likely to bolster one's experience of self-efficacy in these domains (Blustein, 2008; Brown & Brooks, 1985).

Notwithstanding the lack of direct evidence of career interventions enhancing wellbeing, it follows directly from the reviewed literature that interventions assisting individuals to find work contribute indirectly to wellbeing. The routine work of the employment counsellor, such as helping clients with resumes, cover letters, cold calls and employment interviews, is mental health practice insofar as it helps clients obtain work. When these efforts are expanded to include obtaining work that fits the individual (e.g. career development interventions focused on self-awareness, decision-making, values prioritisation), there are greater contributions to wellbeing. Wellbeing outcomes are further enhanced with guidance and counselling activities (e.g. assisting clients with labour market research, informational interviewing, refined decision-making) that assist individuals to find "good" work. Finally, initiatives that develop individuals' capacity to continuously manage their own career development contribute to ensuring ongoing wellbeing. In essence, all career development interventions are wellbeing interventions.

Capacity-building in career development guidance and counselling has traditionally focused on school-aged students. This effort needs to be continued with individuals in the workforce partially because "career literacy" (Magnusson & Redekopp, 2011) can always be reinforced and enhanced, but predominantly because many career management competencies become most meaningful and relevant when the individual is immersed in the environment in which they are needed: i.e. the workplace.

Career development capacity-building directly supports wellbeing. Consider a counselling intervention exploring work activities, one which assists individuals in understanding work stressors as demands to be coped with rather than uncontrollable – but necessary – threats to wellbeing that go along with working life. Perception (appraisal) plays a key role in alleviating problematic stress. Individuals perceiving demands as manageable by their capacity or competence can avoid stress associated with worry about not coping (Hiebert, 1988; Lazarus & Folkman, 1984). This approach suggests that helping workers build skills for recognising, properly framing, and coping with their most important demands will do much to address problematic work stress. Of the work factors identified as contributing to the development of mental health concerns (high job strain, lack of appropriate reward for effort, lack of organisational justice, job insecurity and downsizing, and interpersonal problems and being bullied), most offer some promise of being addressed by skill development that would address coping more effectively with the demands.

Capacity-building efforts enable individuals to be better able to be agents for themselves. If work related stressors (including lack of fit) can be understood as demands, individuals can focus on coping strategies and resources. Coping, expecting to cope, or perceiving that one has capacity to cope with important demands – and most individuals consider work and its role in our lives to be "important" – is a buffer to stress. Individuals with better career management skills are more likely to expect to cope and are therefore less likely to experience stress in reaction to the any instability or transition in their working life. Stress management interventions focused on physiological reactions (e.g. relaxation training, exercise, adequate sleep etc.) have a place in reducing stress, but long-term coping is best achieved by reducing demands or increasing coping ability (Hiebert, 1988). For example, individuals who experience public speaking as a significant source of stress may be best served by job redesign emphasising other competencies, especially those related to strengths. If reducing demands is not optional or desirable, perhaps because redesign is impossible, or because there are benefits, such as salary or promotion, linked to coping, then it makes most sense to improve coping ability (skills) to handle the demand (Hiebert, 1988). In this case, specific training in public speaking might be recommended. Stress is reduced and work-related self-efficacy is strengthened by building skill at public speaking, for example.

Organisational interventions

In seeking the factors and causes of "good work" environments, Lindberg et al.'s (Lindberg et al., 2017) review found "good leadership" (i.e. fair, supportive and empowering leadership) correlated with increased wellbeing and reduced absences due to illness. Transformational leadership (the most studied and supported leadership theory (Barling, 2014)) increased psychological wellbeing and job satisfaction. The question regarding the best ways (e.g. training, mentorship, leadership selection) to obtain or develop leadership remains open.

In two related reviews, Czabała et al. (2011) and Czabała and Charzyńska (2014) reviewed individual, organisational and combined organisational interventions aimed at improving mental health. They found the research focused on five areas: mental health improvement (including maintenance and enhancement, increased job satisfaction (prevention of work overload/burnout, attitude improvements, conflict reduction), job effectiveness improvement, stress reduction and reduction of absenteeism, sick leave and turnover. Generally, interventions (e.g. relaxation, exercise, training) helped lower stress and absenteeism but were far less helpful in increasing job satisfaction and mental health. Of the numerous interventions Czabała et al. (2011) identified, the most effective in each of the five areas above, respectively, were a multi-modal stress management programme; Social Motivational Training (Dupuis & Struthers, 2007); employee problem-solving teams focusing on work effectiveness (a meta-review of workplace interventions addressing depression and anxiety by Joyce et al. (2016) similarly found employee control as a contributor to wellbeing); initiatives developed by staff to respond to possible risks; and stress management training (specifically, biofeedback and muscle relaxation). The most promising approach they found overall was Stress Inoculation Training (Cecil & Forman, 1990) based on Meichenbaum's (Meichenbaum & Deffenbacher, 1988) model. Of interest to counsellors in their review is that counselling employees facing organisational change did not produce significant changes in mental health symptoms or stress indicators yet were rated as the most helpful interventions by employees (Iwi, Watson, Barber, Kimber, & Sharman, 1998).

The findings of Czabała et al. (2011) are helpful, but these authors note that altering the sources of stress will likely have far more powerful effects on stress than the above interventions. This finding holds and is consistent with other reviews that link high job demands with mental illness (Harvey et al., 2017). Czabała et al. also point out that the relatively weak mental health effects found in the literature of various interventions are probably due to using measures of mental illness to detect change. They suggest creating more nuanced measures of mental health to determine change effects. Czabała and Charzyńska (2014) extend this idea by calling for interventions targeting positive mental health rather than only reducing mental health concerns.

Policy interventions

Policy Considerations. "Job satisfaction and positive mood at work both contribute to the productivity of organizations" (Diener & Martin, 2004, p. 10). Given the costs associated with lapses in wellbeing, organisations ought to remind themselves of the cost-effectiveness and potential economic benefit of focusing on and addressing wellbeing concerns. The suggested role of career intervention in wellbeing generally (Robertson, 2013a) and at work (Brown & Brooks, 1985; Herr, 1989) is well-recognized. Most workplace interventions to enhance wellbeing have focused on stress and stressor management. A focus on work stress is supported given the association between work demands and associated stress and the development of mental health concerns (Czabała et al., 2011; Harvey et al., 2017).

A more protective and proactive approach will focus on bettering the work environment. Petrie et al. (2018) developed a five-strategy model organised around concepts of prevention and recovery for individuals working in organisations which includes strategies focused on: (1) designing work to minimise harm; (2) building organisational resilience through good management; (3) enhancing personal resilience; (4) promoting and facilitating early help-seeking; and (5) supporting recovery and return to work. Each of the strategies includes a range of specific interventions to achieve the above outcomes, but as a general policy approach, Petrie et al. emphasise wellbeing-focused organisational policy and an achievable and testable means by which organisations can enhance workplace wellbeing.

From a career development perspective, organisational policy that encourages person-role alignment is an obvious wellbeing effort. Organisational placement and promotion systems typically emphasise organisational need (i.e. people are placed in roles simply because the organisation needs the roles fulfilled), employee skill (i.e. technical competence) and/or employee seniority/tenure in role assignments. Policy that encourages including person-role alignment criteria, especially insofar as it also encourages engaging employees in the discussion of alignment, will support wellbeing.

Additional organisational policy that stimulates employee understanding of, capacity for and involvement in decisions regarding workplace role assignments and demands will enhance wellbeing further. Policy of this nature leads to employee competence in managing their career development and informed control of key decisions that affect their work lives and ultimately, their whole lives.

In terms of public policy and career development, encouraging career development education in developmentally appropriate ways throughout primary and secondary schooling becomes a way of promoting wellbeing and the capacity for individuals to sustain wellbeing. The work of the United Kingdom's recent Gatsby Benchmarks (Boys & Hooley, 2017) provide an example of this kind of policy (e.g. "Every school and college should have an embedded programme of career education and guidance that is known and understood by pupils, parents, teachers, governors and employers" (p. 13)). Australia's "career and skills pathways" approach similarly provides policy direction examples including but going beyond compulsory education to include the role of post-secondary educational institutions, government and employers (Price-Waterhouse-Coopers, 2017).

Conclusion

Although a great deal remains to be clarified about the reciprocal relationships between work and wellbeing, the evidence supports the following, listed in order of contribution to wellbeing:

- Working is better than not working.
- Working in a role that fits one's strengths and proclivities is better than just working.
- Working in a fitting role within a good or decent work environment is better than just working in a fitting role.
- Working in a good and fitting role and having the capacity to adapt to and create change is better than just working in a good and fitting role. In other words, developing career development capacity is a significant wellbeing intervention.

The above findings need to be considered with the caveat that some work can diminish wellbeing and cause mental illness.

Given the above generalisations while recognising that much in the area remains to be understood, it would serve guidance and counselling practitioners, as well as the wellbeing of their clientele, to:

- Develop and use measures by which evidence of mental health / wellbeing changes can be detected rather than using measures of changes in mental illness. Mental illness and mental health may be on different continua, in which case measuring one to detect the other may be futile.

- Collect, analyse and share evidence of wellbeing outcomes arising from career development related guidance and counselling interventions. The career development-work-wellbeing relationships need to be clarified and communicated.
- Continue and strengthen efforts to not only provide career development guidance and counselling services, but to ensure these go well beyond simple work search support or career decision-making, and to include career development capacity enhancement as much as possible.
- Advocate for public policy, organisational/workplace policy and organisational intervention that support decent work that fits employees in which employees have agency and competence in managing their own career development within changing work environments.
- Communicate the wellbeing benefits accruing from career development interventions to other guidance and counselling practitioners so that they better understand the impact of their work.

It may also behoove practitioners to occasionally adopt a perspective of their work and its impact that is wider than typically required by their role descriptions and funding requirements. Remembering the broader aims and far-reaching impact of career development interventions may be energising, providing a boost to their own wellbeing.

Note

1. The goal of this paper is to provide a relevant summary of previously published work with concluding emphasis on future directions and/or areas requiring further consideration. In accordance with accepted procedures for non-systematic reviews (Ferrari, 2015), no methodology is provided in this paper.

Disclosure statement

No potential conflict of interest was reported by the authors.

References

Adler, A. (1958). *What life should mean to you*. New York: Capricorn Books.
Ahola, K., Vuori, J., Toppinen-Tanner, S., Mutanen, P., & Honkonen, T. (2012). Resource-enhancing group intervention against depression at workplace: Who benefits? A randomised controlled study with a 7-month follow-up. *Occupational and Environmental Medicine, 69*(12), 870–876. doi:10.1136/oemed-2011-100450

American Psychiatric Association. (2013). *Diagnostic and statistical manual of mental disorders: DSM-5* (5th ed.). Washington, DC: American Psychiatric Association.

Barling, J. (2014). *The science of leadership: Lessons from research for organizational leaders.* New York: Oxford University Press.

Blustein, D. (2006). *The psychology of working: A new perspective for career development, counseling, and public policy.* Mahwah, NJ: Lawrence Erlbaum Publishers.

Blustein, D. (2008). The role of work in psychological health and well-being: A conceptual, historical, and public policy perspective. *American Psychologist, 63*(4), 228–240. doi:10.1037/0003-066X.63.4.228

Boychuk, C., Lysaght, R., & Stuart, H. (2018). Career decision-making processes of young adults wiht first-episode psychosis. *Qualitative Health Research, 28*(6), 1016–1031. doi:10.1177/1049732318761864

Boys, J., & Hooley, T. (2017). *State of the nation 2017. Careers and enterprise provision in England's schools.* London. Retrieved from https://www.careersandenterprise.co.uk/sites/default/files/uploaded/state_of_the_nation_report_digital.pdf

Brand, J. E. (2015). The far-reaching impact of job loss and unemployment. *Annual Review of Sociology, 41*(1), 359–375. doi:10.1146/annurev-soc-071913-043237

Brown, D., & Brooks, L. (1985). Career counseling as a mental health intervention. *Professional Psychology: Research and Practice, 16*(6), 860–867. doi:10.1037/0735-7028.16.6.860

Caporoso, R. A., & Kiselica, M. S. (2004). Career counseling with clients who have a severe mental illness. *The Career Development Quarterly, 52*(3), 235–245. doi:10.1002/j.2161-0045.2004.tb00645.x

Cecil, M. A., & Forman, S. G. (1990). Effects of stress inoculation training and coworker support groups on teachers' stress. *Journal of School Psychology, 28*(2), 105–118. doi:10.1016/0022-4405(90)90002-O

Czabała, C., & Charzyńska, K. (2014). A systematic review of mental health promotion in the workplace. In C. L. Cooper (Ed.), *Wellbeing: A complete reference guide.* Hoboken, NJ: Wiley Online Library.

Czabała, C., Charzyńska, K., & Mroziak, B. (2011). Psychosocial interventions in workplace mental health promotion: An overview. *Health Promotion International, 26*(suppl_1), i70–i84. doi:10.1093/heapro/dar050

Dawson, C., Veliziotis, M., Pacheco, G., & Webber, D. J. (2015). Is temporary employment a cause or consequence of poor mental health? A panel data analysis. *Social Science & Medicine (1982), 134,* 50–58. doi:10.1016/j.socscimed.2015.04.001

Diener, E., & Martin, E. P. S. (2004). Beyond money: Toward an economy of well-being. *Psychological Science in the Public Interest, 5*(1), 1–31. doi:10.1111/j.0963-7214.2004.00501001.x

Dupuis, R., & Struthers, C. W. (2007). The effects of social motivational training following perceived and actual interpersonal offenses at work. *Journal of Applied Social Psychology, 37*(2), 426–456. doi:10.1111/j.0021-9029.2007.00167.x

Faragher, E. B., Cass, M., & Cooper, C. L. (2005). The relationship between job satisfaction and health: A meta-analysis. *Occupational and Environmental Medicine, 62*(2), 105–112. doi:10.1136/oem.2002.006734

Ferrari, R. (2015). Writing narrative style literature reviews. *Medical Writing, 24*(4), 230–235. doi:10.1179/2047480615Z.000000000329.

Harnois, G., & Gabriel, P. (2000). *Mental health and work: impact, issues and good practices.* Geneva: World Health Organisation/International Labour Organisation.

Harvey, S., Joyce, S., Modini, M., Christensen, H., Bryant, R., Mykletun, A., & Mitchell, P. B. (2013). *Work and depression/anxiety disorders: A systematic review of reviews.* Melbourne. Retrieved from https://www.beyondblue.org.au/docs/default-source/research-project-files/bw0204.pdf?sfvrsn=4

Harvey, S., Modini, M., Joyce, S., Milligan-Saville, J., Tan, L., Mykletun, A., … Mitchell, P. (2017). Can work make you mentally ill? A systematic meta-review of work-related risk factors for common mental health problems. *Occupational and Environmental Medicine, 74*(4), 301–310.

Herr, E. L. (1989). Career development and mental health. *Journal of Career Development, 16*(1), 5–18. doi:10.1177/089484538901600102.

Hiebert, B. (1988). Controlling stress: A conceptual update. *Canadian Journal of Counselling and Psychotherapy / Revue Canadienne de Counseling et de Psychothérapie, 22*(4), 226–241.

Huppert, F. A. (2014). *Interventions and policies to enhance wellbeing* (Vol. VI). Chichester, West Sussex: Wiley Blackwell.

Iwi, D., Watson, J., Barber, P., Kimber, N., & Sharman, G. (1998). The self-reported well-being of employees facing organizational change: Effects of an intervention. *Occupational Medicine (Oxford, England), 48*(6), 361–368. doi:10.1093/occmed/48.6.361

Joyce, S., Modini, M., Christensen, H., Mykletun, A., Bryant, R., Mitchell, P. B., & Harvey, S. B. (2016). Workplace interventions for common mental disorders: A systematic meta-review. *Psychological Medicine, 46*(4), 683–697. doi:10.1017/S0033291715002408

Keyes, C. L. M. (2005). Mental illness and/or mental health? Investigating axioms of the complete state model of health. *Journal of Consulting and Clinical Psychology, 73*(3), 539–548. doi:10.1037/0022-006X.73.3.539

Kossen, C., & McIlveen, P. (2017). Unemployment from the perspective of the psychology of working. *Journal of Career Development.* doi:10.1177/0894845317711043

Layard, R. (2011). *Happiness: Lessons from a new science.* London: Penguin UK.

Lazarus, R. S., & Folkman, S. (1984). *Stress, appraisal, and coping.* New York: Springer.

Lindberg, P., Karlsson, T., Nordlöf, H., Engström, V., & Vingård, E. (2017). *Factors at work promoting mental health and well-being - a systematic litterature review.* Paper presented at the 12th International Conference on Occupational Stress and Health, "Work, Stress and Health 2017: Contemporary Challenges and Opportunities", 7–10 June, 2017, Minneapolis, USA. Retrieved from http://urn.kb.se/resolve?urn=urn:nbn:se:hig:diva-24779

Lindberg, P., & Vingård, E. (2012). Indicators of healthy work environments--a systematic review. *Work (Reading, Mass.), 41* (Suppl 1), 3032–3038.

Luciano, A., & Carpenter-Song, E. A. (2014). A qualitative study of career exploration among young adult men with psychosis and co-occurring substance use disorder. *Journal of Dual Diagnosis, 10*(4), 220–225. doi:10.1080/15504263.2014. 962337

Magnusson, K., & Redekopp, D. (2011). Coherent career practice. *Journal of Employment Counseling, 48*(4), 176–178. doi:10.1002/j.2161-1920.2011.tb01108.x

Marx, K. (1891). *Wage, labor and capital.* Chicago: C.H. Kerr.

McGonagle, A. K., Beatty, J. E., & Joffe, R. (2014). Coaching for workers with chronic illness: Evaluating an intervention. *Journal of Occupational Health Psychology, 19*(3), 385–398. doi:10.1037/a0036601

Meichenbaum, D. H., & Deffenbacher, J. L. (1988). Stress inoculation training. *The Counseling Psychologist, 16*(1), 69–90. doi:10.1177/0011000088161005

Modini, M., Joyce, S., Mykletun, A., Christensen, H., Bryant, R. A., Mitchell, P. B., & Harvey, S. B. (2016). The mental health benefits of employment: Results of a systematic meta-review. *Australasian Psychiatry, 24*(4), 331–336. doi:10.1177/1039856215618523

Parsons, F. (1909). *Choosing a vocation.* Boston: Houghton Mifflin.

Peruniak, G. S. (2010). *A quality of life approach to career development.* Toronto: University of Toronto Press.

Petrie, K., Joyce, S., Tan, L., Henderson, M., Johnson, A., Nguyen, H., … Harvey, S. (2018). A framework to create more mentally healthy workplaces: A viewpoint. *Australian & New Zealand Journal of Psychiatry, 52*(1), 15–23. doi:10.1177/0004867417726174

Praskova, A., Creed, P. A., & Hood, M. (2015). Self-Regulatory processes mediating between career calling and perceived employability and life satisfaction in emerging adults. *Journal of Career Development, 42*(2), 86–101. doi:10.1177/0894845314541517

Price-Waterhouse-Coopers. (2017). *Career and skills pathways: Research into a whole-of-system approach to enhancing lifelong career support mechanisms for all Australians - final Report.* Retrieved from https://www.australianapprenticeships. gov.au/sites/ausapps/files/publication-documents/pwc_career_and_skills_pathways_project_-_main_report.pdf

Robertson, P. J. (2013a). Career guidance and public mental health. *International Journal for Educational and Vocational Guidance, 13*(2), 151–164. doi:10.1007/s10775-013-9246-y

Robertson, P. J. (2013b). The well-being outcomes of career guidance. *British Journal of Guidance & Counselling, 41*(3), 254–266. doi:10.1080/03069885.2013.773959

Sainsbury, R., Irvine, A., Aston, J., Wilson, S., Williams, C., & Sinclair, A. (2008). Mental health and employment.

Salmela-Aro, K., Mutanen, P., & Vuori, J. (2012). Promoting career preparedness and intrinsic work-goal motivation: RCT intervention. *Journal of Vocational Behavior, 80*(1), 67–75. doi:10.1016/j.jvb.2011.07.001

Schumpeter, J. A. (1942). *Capitalism, socialism, and democracy.* New York: Harper & Brothers.

Sun Life Financial. (2012). 2012 *Canadian health index report.* Retrieved from https://cdn.sunlife.com/static/ca/ Learn20and20Plan/Market20insights/Canadian20Health%20index/Canadian_Health_Index_2012_en.pdf

Super, D. E. (1957). *The psychology of careers: An introduction to vocational development* (1st ed). New York and Evanston: Harper & Row.

Thern, E., de Munter, J., Hemmingsson, T., & Rasmussen, F. (2017). Long-term effects of youth unemployment on mental health: Does an economic crisis make a difference? *Journal of Epidemiology & Community Health, 71*, 344–349. doi:10. 1136/jech-2016-208012.

Toppinen-Tanner, S., Böckerman, P., Mutanen, P., Martimo, K.-P., & Vuori, J. (2016). Preventing sickness absence with career management intervention: A randomized controlled field trial. *Journal of Occupational and Environmental Medicine, 58*(12), 1202–1206. doi:10.1097/JOM.0000000000000887

Vuori, J., Toppinen-Tanner, S., & Mutanen, P. (2012). Effects of resource-building group intervention on career management and mental health in work organizations: Randomized controlled field trial. *The Journal of Applied Psychology, 97* (2), 273–286. doi:10.1037/a0025584

Waddell, G., & Burton, A. K. (2006). *Is work good for your health and wellbeing?* London: The Stationery Office.

Warr, P., & Inceoglu, I. (2018). Work orientations, well-being and job content of self-employed and employed professionals. *Work, Employment and Society, 32*(2), 292–311. doi:10.1177/0950017017717684

Westerhof, G. J., & Keyes, C. L. M. (2010). Mental illness and mental health: The two continua model across the lifespan. *Journal of Adult Development, 17*(2), 110–119. doi:10.1007/s10804-009-9082-y

World Health Organization. (2017). *Depression and other common mental disorders: Global health estimates.* Geneva: World Health Organization.

Happiness is not a luxury: interview with Ed Diener

Anuradha J. Bakshi

Introduction

Edward Diener is the Alumni Distinguished Professor of Psychology (Emeritus) at the University of Illinois, USA. He is also a professor of psychology at two other U.S. universities: University of Virginia and the University of Utah. As an academic, he began working on happiness and wellbeing in 1980. With around 400 publications and a citation count of 180,000, he is one of the most cited psychologists in the world. As most of his work is on happiness and wellbeing, he has earned the moniker Dr Happiness. His many honours and recognitions include the Distinguished Scientific Contribution Award from the American Psychological Association and the Distinguished Quality-of-Life Researcher Award from the International Society of Quality of Life Studies. He has served as a president of the International Society for Quality of Life Studies, the International Positive Psychology Association, and the Society for Personality and Social Psychology. For a substantial part of his career he was the editor of *Journal of Personality and Social Psychology* and a senior scientist with Gallup. He is also the founding editor of *Perspectives on Psychological Science*, and a co-founding editor of *Journal of Happiness Studies*.

Anuradha Bakshi is one of the two Co-Editors of the British Journal of Guidance and Counselling. She heads the Department of Human Development at Nirmala Niketan College of Home Science, University of Mumbai, India. Positive psychology is one of her interest areas and she and her students have conducted research in spirituality, religiousness, forgiveness and altruism.

The following interview was conducted through an exchange of emails between Anuradha Bakshi and Ed Diener.

Interview

Anu: Thank you so much for agreeing to be interviewed for our special issue on Happiness and Wellbeing. I have mulled over some questions for the interview. You have published so extensively and so impactfully, that I do hope you do not find the questions overly basic. Also, I have written a few more questions than perhaps you may wish to answer but I thought it important to pen down a range of ideas such that you could have a choice.

Ed: They are great questions, but could easily take me a day or two to answer. Would it be possible just to give me a few of them to answer, the most important ones?

Anu: Yes, I did send you quite a number of questions, and as you are so well published the answers to some or many of the questions are already in your publications. My intention here is to capture and spotlight some of your key perspectives on wellbeing in our symposium issue on Happiness and Wellbeing. Simultaneously, I was hoping to have our readers also find some new content from you in case any of my questions happen to be ones that you have not specifically answered earlier. Let's start with a smaller selection of questions that can meet this dual aim.

You have dedicated a lifetime of effort in theorising about and researching well-being. Have there been any surprises along the way? Where you had an insight or came across a finding which previously you had not expected?

Ed: I have encountered several surprises in exploring why some people are happier than others. One of the big surprises for me is that the country where people live can make a huge difference. It is not just whether a person is rich or poor or educated or uneducated, but the societal characteristics of where they live that can make a huge difference in their subjective well-being. Factors such as economic development, trust and respect, and health matter a great deal for well-being, and so we must be concerned about societal policies, not just making individuals happier. Another surprise is that although income matters quite a bit for life satisfaction, after a certain point of income it no longer makes much difference. Thus, coming out of poverty is important for happiness, but going from well-off to rich does not boost it that much. A third surprise here is that although income and money matter for life satisfaction, for enjoying life social relationships are a key, regardless of whether one is rich or poor.

One last surprise that I learned very early in my research is that happiness and unhappiness are not opposites – having positive affect and enjoying life is not strictly the opposite of negative affect and worry and sadness. The two have some independence. Thus, in order to make people truly "happy" we cannot just work on ridding them of negative affect; we must create enjoyment of work, love, and life as well.

Anu: Your first surprise ties in so well with a systems view. Policies can help create contexts that (in turn) help engender wellbeing among individuals. This also shifts at least part of the responsibility to macro-level organisations including governments.

Your second surprise about income: I have observed that so long as an individual or a family is not struggling with destitution, higher and higher income can well generate stress and reduced enjoyment. For one, the individual or family may wish to push their standard of living higher in tandem with increased income, and aim for a competitively expensive lifestyle which puts a strain on their finances and their relationships. And as you pointed out, these very same social relationships are crucial for enjoying life. Your work on social comparison is very relevant. You have contended that with rising income, individuals on average are not happier because they have raised the bar for themselves, and now mark their success against a higher standard. Plus, I think status maintenance, wealth management, ensuring safety of material acquisitions may each necessitate extensive investment of resources and potentially cause anxiety, obstructing any elevation in happiness. I also found it interesting that you have some suggestive evidence that wellbeing is less likely to be enhanced if money is spent on material objects; on the other hand, if money is spent on others and on experiences, subjective wellbeing is enhanced.

Your fourth surprise links to some other ideas that you have shared in your writing. The absence of negative affect is not good enough, and we must expend effort to experience positive affect. Yet, we don't just become happy once and for all. Happiness is not a destination; continual engagement is needed to sustain positive affect. I would like to quote you here:

> One thing that is quite clear to me is that happiness is a process, not a place. No set of good circumstances will guarantee happiness. Although such circumstances (a good job, a good spouse, and so forth) are helpful, happiness requires fresh involvement with new activities and goals – even perfect life circumstances will not create happiness. (Diener, 2008, p. 6)

There is such a range of variables that have been linked to wellbeing. Can you compare the contribution of different factors to wellbeing – such as mindfulness, self-esteem, self-determination, positive affect, optimism, religiousness, spirituality, personality, social support, income, health, quality of life, and culture?

Ed: We find some factors are associated with well-being across cultures, but others seem more culture-specific. For example, social support seems important everywhere we look in the world – across diverse cultures. But self-esteem, which I once thought would be a universal and necessary predictor of happiness, seems like it is a strong correlate of life satisfaction in only some cultures. My daughter, Marissa Diener, and I found that women in southern India, for example, did not show much association between "self-esteem" and life satisfaction – being satisfied with themselves was simply not the way they thought about their lives. In contrast, having people who you trust, who respect you, and who you can count on in an emergency appear to be universals that raise people's positive feelings everywhere in the world.

Anu: I notice that your findings are very similar with regard to many other variables including religion. Using Gallup World Poll data with a sizable sample of 455,104 individuals from 154 countries, you have reported (Diener, Tay, & Myers, 2011) that the relationship between religiousness and wellbeing is contingent on societal characteristics. So, regardless of which religion an individual professed to and practised, religious individuals had higher subjective wellbeing than less religious individuals in countries characterised by difficult circumstances (such as going hungry, unsafe neighbourhoods, low educational attainment and low life expectancy); whereas in affluent countries the religious and less religious had comparable levels of wellbeing. You also uncovered a person-culture fit effect such that religious individuals in religious countries had higher wellbeing but religious individuals in non-religious countries did not.

We appear to be circling back to a systems view. I would like you to elaborate on the salience of contexts. To what extent is wellbeing (i.e. markers, strategies, experience, outcomes) context-dependent? Can you give us some (more) examples?

Ed: Let me emphasise here some of the individual factors that make different people happy for different reasons. A lot of the research on well-being is based on averages. So, when researchers say that married people are happier, or people with money are happier, or that people with children are no happier than others, they are always talking about mean averages. The average might not apply to specific individuals that well, and this is so important to convey to people. People with children are on average no happier or less happy than people without children, correct. BUT this might not apply to you – you may love children and be much happier with them, or you might not enjoy kids, and they will just be a burden. Similarly, some people are happier because they have a great marriage, and others are miserable in marriage. This individuality is so important to convey to people because it has a lot to do with what they pursue to be happy. What type of person are you? What type of partner will truly make you happier in the long run, if any? Too many of our past recommendations were based on averages, without fine-tuning them for the individual.

Anu: This is so helpful. Another field in which the one-size-fits-all mantra is no longer current. Instead, recognition that wellbeing is nuanced.

I have a couple of questions specifically related to the journal's aims and scope. As the British Journal of Guidance and Counselling particularly focuses on career development, what conclusions would you like to especially emphasise with regard to the relationship of happiness and well-being to work and career?

Ed: So often young people in choosing a career select one that is of high income and high status. These characteristics are fine, of course, but not the most important ones in creating a happy life, even though some moderate income can be essential for subjective well-being. Instead, I recommend a job that includes the kinds of things you love doing, and which you are good at. I love analysing data, teaching students, and collecting data. And so I absolutely love my job, even if I do not make as much money as in some other professions. Of course, since we spend 40 or so hours a week in work, it is what we do most with our waking time. And thus, if we are happy at work we are happy the majority of our time. Contrariwise, people sometimes choose work that will make money but which they will not enjoy, thinking that they can make up for disliking their work because in their nonwork-time they can then afford to do fun things such as travel, have a nice home, and so forth. But how can one be unhappy 40 hours a week to be happy the other hours – it simply does not work that well. I am a very happy person because I have work I love. After all, most people find leisure pursuits they love.

Anu: You have put the role of happiness vis-à-vis work and leisure in perspective. Your argument is simple and yet very cogent. Those of us in full-time positions do spend most of our waking time engaged in work: Are we doing what we love in all those hours that eventually add up to most of our lifetime?

What (else) would you like to bring to the attention of counsellors, psychotherapists and careers workers? The decades of work that you and your colleagues have completed on happiness and well-being, how can it benefit the work of counsellors, psychotherapists and careers workers?

Ed: One of our most important findings over the past decade is that subjective well-being is not just pleasant and feels good, it is good for people. People high in well-being have better health and long-evity, and they have better marriages and more supportive friends. Sad and angry people don't just feel bad, but they also do not function as well on average in social relationships or at work. Happy people tend to do more things to keep themselves healthy. Some critics warned that happy people are self-centered and don't care about societal problems. But we might see that although they may worry a bit less about such societal issues, they in fact do somewhat more activities to remedy problems. Happy people are more likely to get involved with activities that may help solve the problems, and they tend to be better citizens. Thus, happiness is not a luxury – societal leaders need to understand that having a happy society is not just feel-good, but is actually helpful in improving the overall quality of life.

Anu: Happiness is good for individuals, happiness is good for groups. This appears to break the myth that angst is a necessary catalyst for exceptional achievements or contributions. Yet, in your writing you have questioned whether there is an optimum level of happiness, and whether there needs to be an optimal balance of positive and negative affect. The emotional regulation literature is very rich and can provide some answers. I'm also wondering whether we need to study types of happiness. Aristotle appears to have made a headstart here, and psychology distinguishes between hedonic (related to pleasure) versus eudaimonic (related to self-actualisation) wellbeing. Subjective wellbeing, of course, is not identical to either as it refers to individuals making an evaluation of their own quality of life themselves. I wish to mention that congruent with an Eastern perspective, I find that when I communicate best wishes to another I am less likely to wish them happiness, and I am more likely to wish them joyous fulfilment. Perhaps we need persons to evaluate both for their own selves: a fun type of happiness, and a fulfilment type of happiness. Would you agree?

Ed: People may seek both a bit of fun, and long-term sustainable fulfilling happiness. But they may not be either-or. One might find the fulfilling happiness to be pleasant, as I do, and the fun things to be fulfilling, often because they are done with people I care about. Of course, the two sometimes do converge, but best when our activities are both! Then you are really blessed.

Anu: A fulfilling happiness that is simultaneously fun, and fun types of happiness that are also fulfill-ing. I like that! Even more so the idea that thriving entails bringing together rather than separating fulfilment and fun. Thank you for endorsing the value of fun, and by implication its companion – playfulness.

Ed, is there anything else that you would like to say as we close this interview?

Ed: There are two important findings that everyone must understand now, as these scientific findings about happiness can change everything. I am not exaggerating; the findings can change lives and policies! First, we now know that happiness can be raised – by interventions with individuals, with programs designed to teach people skills that will improve their happiness and quality of life. Also, wellbeing can be raised at the societal level with the policies that are adopted – for example, in creating more green space in cities, reducing air pollution, and in reducing commuting. We now know that community and societal characteristics can have a huge influence on subjective well-being. Second, we know that raising happiness is not just a good idea because it feels better to be happy. I wish to reiterate that wellbeing has huge effects on the other things we care about most deeply – health and longevity, work performance, resilience to stress, supportive social relationships, and good citizenship. The effects of well-being are large and broad and these are factors that are important to leaders and politicians. Even if they do not highly value happiness, these characteristics are important to them. So, the findings on the benefits of happiness are truly "game-changers!" Happier societies are healthier, more supportive, and creative. It is our mission to bring this knowledge to the world! The ways we know will improve happiness will also improve communities, societies, and the world!

Anu: Thank you so much, Ed. In concluding, you have drawn attention to the modifiability or plasticity of happiness both at individual and societal levels. And you have offered happiness as a tool for optimising development, again at both individual and group levels. Once again, we remind our readers, "happiness is not a luxury", it's a necessity for living healthily, productively, resiliently, and peacefully as individuals or groups.

References

Diener, E. (2008). One happy autobiography. In R. V. Levine, A. Rodriques, & L. Zelezny (Eds.), *Journeys in social psychology: Looking back to inspire the future* (pp. 1–18). New York, NY: Taylor & Francis.

Diener, E., Tay, L., & Myers, D. G. (2011). The religion paradox: If religion makes people happy, why are so many dropping out? *Journal of Personality and Social Psychology, 101*, 1278–1290.

Index

Note: **Bold** page numbers refer to tables and *italic* page numbers refer to figures.

Printed and bound by CPI Group (UK) Ltd, Croydon, CR0 4YY

18/10/2024

01776250-0014